MEGA-DAMS IN WORLD LITERATURE

MEGA-DAMS IN WORLD LITERATURE

MEGA-DAMS IN WORLD LITERATURE

Literary Responses to Twentieth-Century Dam Building

Margaret Ziolkowski

UNIVERSITY OF WYOMING PRESS
Laramie

© 2024 by University Press of Colorado

Published by University of Wyoming Press
An imprint of University Press of Colorado
1580 North Logan Street, Suite 660
PMB 39883
Denver, Colorado 80203-1942

All rights reserved
Printed in the United States of America

 The University Press of Colorado is a proud member of
the Association of University Presses.

The University Press of Colorado is a cooperative publishing enterprise supported, in part, by Adams State University, Colorado State University, Fort Lewis College, Metropolitan State University of Denver, University of Alaska Fairbanks, University of Colorado, University of Northern Colorado, University of Wyoming, Utah State University, and Western Colorado University.

∞ This paper meets the requirements of the ANSI/NISO Z39.48-1992 (Permanence of Paper).

ISBN: 978-1-64642-595-2 (hardcover)
ISBN: 978-1-64642-596-9 (paperback)
ISBN: 978-1-64642-597-6 (ebook)
https://doi.org/10.5876/9781646425976

Library of Congress Cataloging-in-Publication Data can be found online.

Cover art by Lara Thurston

In memory of my father and mother,

Theodore and Yetta Ziolkowski

Contents

List of Figures   ix

Preface and Acknowledgments   xi

Maps   xiii

1. Introduction   3

2. The High Modernist Heyday of Mega-Dam Construction   27

3. Displacement and Alienation of Peoples Worldwide   70

4. Contaminated Water, Disappearing Fish, and Deadly Sediment   114

5. Dam Failures, Real, Imagined, and Ecotage-Inspired   145

Conclusion   172

Notes   177

Bibliography   193

Index   201

Contents

List of Figures ix

Preface and Acknowledgements xi

Maps xiii

1. Introduction

2. The High Modernist Hubris of Mega-Dam Construction 27

3. Displacement and Alienation of Peoples Worldwide 70

4. Contaminated Water, Disappearing Fish, and Deadly Sediment 114

5. Dam Failures: Real, Imagined, and Propaganda-Inspired 145

Conclusion 172

Notes 177

Bibliography 193

Index 201

Figures

0.1.	Dnieper and Volga Rivers	xiii
0.2.	Angara River	xiv
0.3.	Columbia, Colorado, and Missouri Rivers	xv
0.4.	Tennessee River	xv
0.5.	Nile River	xvi
0.6.	Yangtze River	xvii
1.1.	*President Franklin D. Roosevelt at the Opening of Hoover Dam*	4
2.1.	Pyotr Ivanovich Kotov, *Maxim Gorky at the Construction of the Dnieper Hydroelectric Station*	33
2.2.	Max Alpert, *Destroyed Dnieper Hydro Electric Power Station*	37
2.3.	P. Hughes, *View of Hoover Dam and Part of Lake Mead, Nevada-Arizona Border*	49
2.4.	*Grand Coulee Dam on the Columbia River, Washington*	58
2.5.	Al Aumuller, *Woody Guthrie, Half-Length Portrait, Facing Slightly Left, Holding Guitar*	59
3.1.	*Gamal Abdul Nasser with Nikita Kruschev, during the Ceremony of the Divert of the Nile at Aswan High Dam*	93
3.2.	Christoph Filmkössl, *Three Gorges Dam*	111
5.1.	André Payan-Passeron, *Glen Canyon Dam*	160

Preface and Acknowledgments

While working on my last book, *Rivers in Russian Literature*, I found myself increasingly fascinated by the impact of large hydroelectric dams on people and the environment. In investigating this topic, I discovered a surprising number of literary works related to dams worldwide that deal with or touch on this topic in a significant manner. I found reading them both informative and moving and decided to share what some might regard as a rather offbeat literary interest.

I had much encouragement and help in working on this project and developing the manuscript. Before they passed away, both my father and my mother, Theodore and Yetta Ziolkowski, provided me with the support they always did when I pursued my intellectual interests. My husband, Robert Thurston, patiently read and commented on every word of the manuscript at least once. My son and daughter, Alexander and Lara Thurston, were uniformly encouraging. Lara, a skilled graphic artist, designed all the maps in the book according to my specifications and her expertise. My good friend and historian Judith Zinnser read several chapters and offered valuable

https://doi.org/10.5876/9781646425976.c000

xii   PREFACE AND ACKNOWLEDGMENTS

perspective drawn from her work at the United Nations. The interlibrary loan staff at Miami University helped me locate several rather obscure books. The director of Miami's Interactive Language Resource Center, Daniel Meyers, once again provided the formatting aid I am often in great need of as someone who is irredeemably technologically challenged. I am also grateful to the anonymous readers of the manuscript who were very helpful in outlining how I might improve both the content and style of my work. Finally, I much appreciate the unfailing support provided by University of Wyoming Press acquisitions editor Robert Ramaswamy, who was easily the most encouraging editor it has ever been my good fortune to encounter. I express my gratitude to all these people for their support and assistance. Any errors in the manuscript are of course my own.

Maps

FIGURE 0.1. Dnieper and Volga Rivers. Map by Lara Thurston.

FIGURE 0.2. Angara River. Map by Lara Thurston.

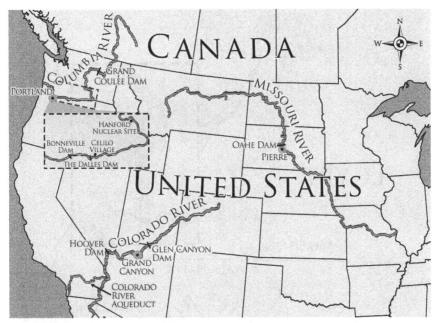

FIGURE 0.3. Columbia, Colorado, and Missouri Rivers. Map by Lara Thurston.

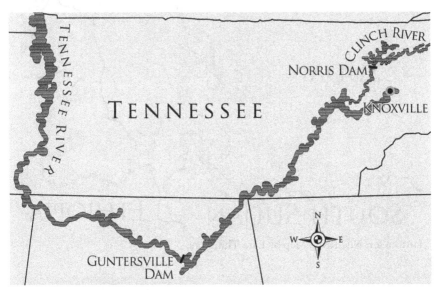

FIGURE 0.4. Tennessee River. Map by Lara Thurston.

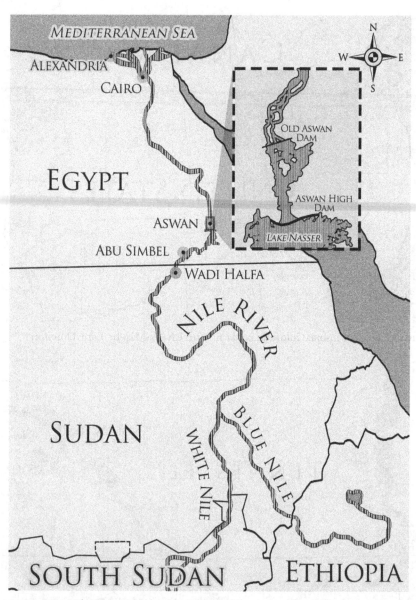

FIGURE 0.5. Nile River. Map by Lara Thurston.

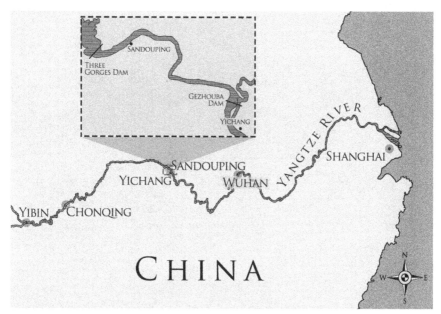

FIGURE 0.6. Yangtze River. Map by Lara Thurston.

MEGA-DAMS IN WORLD LITERATURE

MEGA-DAMS IN WORLD LITERATURE

1

Introduction

"I came, I saw, and I was conquered," declared Franklin Delano Roosevelt in 1935 at the dedication of Hoover (then Boulder) Dam, voicing in his paraphrase of Julius Caesar's widely known exultant exclamation a belief that large hydroelectric dams embodied a victory of the most cherished modern values and aspirations (figure 1.1).[1] Such dams, sometimes termed megadams, were a quintessential physical and cultural feature of the twentieth century and have remained so into the twenty-first century. Writers from a variety of countries have portrayed their construction and environmental, social, and political impact in numerous literary works. This literary representation—which engages readers in a range of important themes related to dams, both directly and subtly—is the central topic of this study.

Initially and at times still hailed as icons of modernism and triumphs of development, throughout the twentieth century and into the new millennium, big dams were and often still constitute objects of desire, envy, and emulation across the world. Since their inception, they have borne important political implications of varying stripes. As structures sought after by

---

https://doi.org/10.5876/9781646425976.c001

FIGURE 1.1. *President Franklin D. Roosevelt at the Opening of Hoover Dam*, photograph, September 30, 1935. Franklin D. Roosevelt Library, http://www.fdrlibrary.marist.edu/daybyday/resource/september-1935-5/.

eager engineers and often greedy politicians alike, they have reinforced ideologies by improving navigation, helping prevent disastrous flooding, and enabling a financially rewarding and broader use of irrigation. They have transcended the particularity of political systems in their obvious visual and societal appeal. Democratic, socialist, communist, authoritarian, developed, developing, underdeveloped—no matter the nature and contours of its government and society, every country that could afford dams or borrow funds to build them had long craved them (and still do), and the bigger, the better. A large dam constitutes an imposing national status symbol, an indication that a country has truly arrived on the world scene or, as Bret Benjamin puts it, acquired one of the critical "fetish objects" of nationalism.[2] Literature, both fiction and poetry, helps underscore this symbolic importance, giving voice to perceived engineering, social, and political victories.

The evolving perspectives of large dams that took place during the course of the twentieth century were remarkable; an understanding of the dilemmas, social and natural, presented by dams is essential to comprehend their importance in literature. This chapter provides an overview of such problems, and subsequent chapters will expand on these issues in relation to the literary works discussed. As the twentieth century progressed, the initial unadulterated enthusiasm such as Roosevelt's comment was tempered by recognition of the immense social and environmental costs of dams, and literature often shifted from appreciation and adulation to reservations about and even hostility toward mega-dams. At every point along this complex path, poets and writers of fiction played a significant cultural role—initially as key propagandists or cheerleaders and later as influential critics and denunciators—in raising public consciousness about the virtues and vices of structures like the American Hoover, Glen Canyon, and Grand Coulee Dams, the Egyptian Aswan High Dam, the former Soviet Union's Dneprostroi and Bratsk, various Indian dams, and China's Three Gorges Dam. Readers who might not tackle scientific articles and books on dams and their consequences may be attracted to novels that touch on the same topics. This means that the subject of dams in all its complexity would reach a larger audience and acquire a more humanized form. A comparative analysis of the literary treatment of dams over time in American, Russian, Ukrainian, Chinese, Indian, and Egyptian writings provides important insights into the cultural apprehension of the benefits of development and industrialization, as well as the subsequent understanding of the deleterious impact of big-dam construction on disadvantaged populations, the environmental damage they have wrought, and the multiple possibilities for corruption and fears of terrorism that accompany large-scale hydroelectric projects. Novels and poems about dams and their construction offer cultural snapshots of technological and political progress and settler colonialism construed as both dream and nightmare. They demonstrate the central role literature can play in expressing and influencing a wide range of popular and political views of mega-dams and provide, for example, a broader corollary to the visual images discussed by Donald C. Jackson in *Pastoral and Monumental: Dams, Postcards, and the American Landscape* (2013). They personalize the impact of dams in all their complexity in a way scientific treatments cannot do.

An appreciation of the potential importance of dams, small and large, goes back thousands of years. Literary works afford some understanding of this long-standing attitude, of the consistent and constant human desire to

6   INTRODUCTION

subordinate nature to human inclinations and the belief that such a desire is divinely endorsed or politically justified. The human longing to control water and waterways has, as in so many other instances involving nature, often included a combination of deprecation of, even contempt for, nature and a desire to dominate it. With hydroelectric dams, such feelings became particularly acute.

By the nineteenth century, an attitude toward nature marked by barely concealed contempt and an inclination toward self-serving usage was firmly in place in many circles. An eagerness to improve upon navigation in early nineteenth-century Germany, for example, led to a program of "rectification" of the Rhine (*Rheinkorrektur*), the literal straightening out of the river, or correction, to facilitate shipping; such processes would be replicated, for example, with the Mississippi River. The United States Bureau of Reclamation, founded at the beginning of the twentieth century, was intended to manage water resources in the American West, particularly as far as irrigation was concerned. The very terms *rectification* and *reclamation* reflect a sense that nature calls for remediation or exploitation and that it is a human right, perhaps even a divinely, biblically enjoined obligation, to engage in this water-obsessed process. As David Owen comments, "In the 1920s, 'conserving' river water meant extracting as much profit from it as possible before it flowed into the sea."[3] Not, one might note, saving or protecting as much of it as possible, in today's conventional meaning of conservation.

As sentient, intentional beings, humans frequently considered themselves superior to unconscious, wild, and dangerous nature. Humans had plans and constructive ideas. Nature, in contrast, often seemed careless and unimaginative and was implicitly or explicitly an object of condescension. A character in Oscar Wilde's dialogue "The Decay of Lying" disparages "nature's lack of design, her curious crudities, her extraordinary monotony, her absolutely unfinished condition."[4] A harsh denunciation indeed. Nature, many educated nineteenth- and twentieth-century observers thought, was often outright wasteful in its unconscious and correspondingly insensitive ways. Rivers afforded a prime and very visible example of such waste. An increasing awareness of the possibilities hydroelectricity afforded only strengthened this perception.

At the turn of the twentieth century, the Indian civil engineer and statesman Mokshagundam Visvesvaraya supposedly declared on seeing the

Introduction   7

impressive Jog Falls on the Sharavati River in western India: "What a waste of energy."[5] A few years later, standing at Owens Falls on Lake Victoria, Winston Churchill was likewise driven to exclaim "so much power running to waste . . . such a lever to control the natural forces of Africa ungripped."[6] Implicitly, an assumption of a colonialist imperative, with the notion of settler colonialism lurking in the background, underwrites this statement. In his 1935 address, Roosevelt declared: "The mighty waters of the Colorado were running unused to the sea. Today we translate them into a great national possession."[7] Again there is an emphasis on the rights of nations. Several decades later, the Canadian premier of Quebec, Robert Bourassa, lamented that "Quebec is a mighty hydroelectric plant in the bud, and every day millions of potential kilowatt-hours flow downhill and out to sea. What a waste."[8] Human beings could and should, often implicitly as a moral directive, address this perceived shortcoming on ignorant nature's part by seizing control of natural forces. Settler colonialism embodied this directive on a large and obvious scale.

It was only a short cultural step from a perception of nature as wasteful to an assumption of the right to dominate and control nature, to exercise what Oswald Spengler called "the Faustian technics, which . . . thrusts itself upon Nature, with the firm resolve to *be its master*."[9] As Lewis Mumford commented a few years later, "The dream of conquering nature is one of the oldest that has flowed and ebbed in man's mind. Each great epoch in human history in which this will has found a positive outlet marks a rise in human culture and a permanent contribution to man's security and well-being."[10] The Enlightenment era in particular had added force to such ideals, feeding the dreams of later thinkers like Spengler and Mumford. Subordinating nature to human needs, in other words, leads to growing wealth and eminently deserved happiness. A vehement late nineteenth-century exponent of this attitude was the industrialist Andrew Carnegie, who asserted in the 1880s: "Man is ever getting Nature to work more and more for him . . . Ever obedient, ever untiring, ever ready, she grows more responsive and willing in proportion as her lord makes more demands upon her."[11] Note the gendering of nature as female, as a compliant object of male attention, and the casting of the male in question as a lord. In 1941, David E. Lilienthal, one-time chair of the Tennessee Valley Authority and author of the immensely influential tome *TVA: Democracy on the March*, expressed similar thoughts at

8   INTRODUCTION

the beginning of his text when he spoke of "a wandering and inconstant river now become a chain of broad and lovely lakes which people enjoy, and on which they can depend, in all seasons, for the movement of the barges of commerce that now nourish their business enterprises. It is a story of how waters once wasted and destructive have been controlled and now work, night and day, creating electric energy to lighten the burden of human drudgery."[12] Implicitly, nature is a servant of human beings, if not an outright slave.

At the Twenty-second Congress of the Communist Party of the Soviet Union in 1961, Nikita Khrushchev agreed with this forceful and human-centered assessment, giving it an explicitly socialist coloring: "Our party will succeed in saving man from the influence of the elements, in making him the master of nature."[13] As had begun even earlier because of Western fears of socialists and socialist conviction regarding the evils of capitalism since the inception of the Soviet Union, competition between the United States and the Soviet Union during the Cold War brought the issue of human control over nature to the forefront. Even later, after the possible detrimental effects of large dams had begun to be recognized at a 1991 conference of the International Commission on Large Dams, Otto Hittmaier, former president of the Austrian Academy of Sciences, continued to argue vigorously for the benefits of large dams and asserted: "Man's first duty is to his species. We should obey the biblical command to go forth and subdue the Earth."[14] As the biblical reference (see Genesis 1:28) suggests, the desire to control nature and the religious conviction of the right to do so go back to the earliest civilizations. Further, such a desire was by no means limited to particular types of political systems. As Murray Feshbach and Alfred Friendly observe, a "view of nature as a cornucopia to be pillaged" does not observe political boundaries.[15] Despite religious appreciation at times for nature's beauty, the idea that nature should serve humanity has always been driven by a range of religious and political ideologies. At every step of the way, literature has provided support for these ideologies and designs, helping convince large numbers of people that human beings should by right seize control of nature.

The desire to dominate nature could take various metaphorical forms. As Carnegie's comments suggest, nature could be imaginatively construed as a potential (female) laborer on behalf of (male) humans. This thought became a dominant refrain in the twentieth century. In 1962, Allen H. Cullen, author of the suggestively titled *Rivers in Harness*, emphatically stated: "All dams . . . serve the same basic purpose: to help man, to work for him, to aid him in the mighty job of conquering his environment."[16] On the other side of the Cold War world, in a very different political context but a similar symbolic vein, Vladimir Sinedubsky, writing about Soviet hydroelectric dams, declared: "The Angara [a Siberian river] hydro-system . . . will turn the wayward daughter of Lake Baikal [the Angara River is the only river that flows out of Lake Baikal] into a diligent labourer."[17] Thus, in very different political contexts, the notion that nature should submit to and work for human beings is a consistent one.

An equally common and logical belief deriving from the conviction that human beings should dominate nature's existence is that nature may emerge as an enemy against which it is necessary and proper to wage war; nature might selfishly not wish to be dominated, but it should be—its potential agency subsumed to human desires, no matter the aggression and force necessary. Martial analogies were frequently employed in discussions of rectification of the Rhine.[18] In France after World War II, hydroelectric development of the Rhône was represented as an "epic 'battle against nature.'"[19] Judith Shapiro mentions the extensive use of military imagery in regard to the Chinese communist attempt to subordinate nature.[20] Paul R. Josephson comments on similar linguistic usage in Brazil in discussions of development of the Amazon region and in the Soviet Union in regard to Siberia.[21] Europe, Asia, Africa, the Americas—it did not matter. Nature was an enemy to defeat; indeed, nature deserved to be overcome by human beings. Nature might possess agency, but it stood in need of human management and even suppression.

Human beings have been trying to dominate nature in myriad ways for eons. What gave particular ideological impetus to this desire starting in the late nineteenth century and continuing throughout the twentieth century was the growth of attitudes exemplifying what has been termed *high modernism*. Drawing on David Harvey's work on postmodernity, James C. Scott defines high modernism as "a strong, one might even say muscle-bound, version of the self-confidence about scientific and technical progress, the expansion of production, the growing satisfaction of human needs, the mastery of nature (including human nature), and, above all, the rational design of social order commensurate with the scientific understanding of natural laws."[22]

10   INTRODUCTION

In other words, modernity became an assumption of human superiority on steroids, a sense of warranted mastery of docile or resistant nature. Any aspirations to agency on nature's part evoked disdain and a conviction that such desires should be overcome by human beings.

Scott and others have stressed that high modernism recognizes no ideological boundaries.[23] The Cold War bore out the accuracy of this observation. Scholars have also emphasized the deep links between modernization discourse, with its preoccupation with reason, science, and domination of nature, and the legacy of the Enlightenment, with its unrelenting emphasis on rational behavior.[24] In turn, as Hittmaier's comment above suggests, the Enlightenment's instrumentalist attitude toward nature may have had "ideological taproots in the Christian doctrine of dominion"[25]—a doctrine, one might note, far from unique to Christianity. Many other religions have aspired to dominate nature, and it often remains very difficult to separate religion and politics. As for the relationship between mega-dams in particular and modernization, in her comparative study of the Volga and the Mississippi, Dorothy Zeisler-Vralsted comments on the centrality dams assumed in the modernization project: "Building monumental dams became the currency of modernization."[26] To modernize meant to build a mega-dam in one's own nation, preferably more than one.

High modernism transcended nationalism and went hand in hand with a preoccupation with agricultural, industrial, and social development; the glorification of technological advances; and the assumed concomitant physical and social role nature played as a kind of handmaiden to technology in the improvement of the human condition that would occur.[27] Dams, it was thought, could greatly facilitate both agricultural and industrial development and, by permitting a higher standard of living, lead to a more broadly engaged and sophisticated society—indeed, to superior people. Such people might be capitalist or socialist, but they would share a capacity for domination of and superiority to thoughtless nature. Hence Vladimir Lenin's famous assertion that communism is Soviet power plus electrification of the entire country, an electrification implicitly linked to hydroelectric potential. Hence also the subtitle of Lilienthal's treatise on the Tennessee Valley Authority (TVA): *Democracy on the March*. As David Ekbladh has pointed out, the TVA "was a grand synecdoche, standing for a wider liberal approach to economic and social development both domestically and internationally."[28]

On the other side of the world, in a very different political context, Soviets also believed in the unmitigated virtues of development. It was long assumed that development could only bring good. The uncomfortable and generally unacknowledged truth that the pursuit of development often provided cover for rapacious neo-colonialism—both domestic and foreign, socialist and democratic—was unrecognized or studiously ignored. As Steven Hawley puts it, "Water . . . began flowing uphill toward money."[29] In this ambiguous process, literature played an important role in bolstering the supposedly indubitable benefits of development. Rivers and the land were waiting for human intervention, and literature could describe this process in a manner unlike any other.

Hydroelectric dams ostensibly provided a key and highly visible piece of evidence for the virtues of development. Sanjeev Khagram argues that big dams were "socially constructed during the twentieth century as premier development activities and symbols."[30] Worldwide, this led to the entrenchment of powerful dam-building bureaucracies operating under the auspices of a wide range of political and financial systems. Institutions like the World Bank were eager to finance dam construction, and the detrimental social and environmental impacts that might follow in the wake of big dams were long overlooked. As Gilbert Rist describes it, this process was driven by "the idea of a natural history of humanity: namely that the *'development' of societies, knowledge and wealth corresponds to a 'natural' principle.*"[31] In other words, development is manifestly righteous, ordained by a wide variety of political and religious thinking. Blanket assumptions of such attitudes—a justification for settler colonialism, among other processes—fed into an uncritical acceptance that dams were an uncontested good. After all, look what dams could do. Writers of various political persuasions were happy to describe the rapture dams engendered, no matter whether they were built in capitalist or communist, authoritarian or democratic societies.

The creation of the United Nations and the World Bank took place at the end of World War II. The UN, it was hoped, would help prevent future wars and act as a more effective institution than had the League of Nations. Over time, the UN shifted from an emphasis on peacekeeping to a focus on economic and social development, from idealism to pragmatism. The World Bank played an increasingly important role in this process, and the notion of assistance gave way to what was often a pretense of cooperation, despite

12   INTRODUCTION

the fact that the real division of labor was between Washington-based international financial institutions and UN agencies.[32] Such an approach did not always have happy results. By the end of the twentieth century, some countries in the Global South were convinced that there was too much interference stemming from the North, too much exploitation, much of it engendered by haughty and self-centered colonialist practices. With new environmental concerns, a new kind of division in thinking arose: "North-South divisions led to a new conceptualization. Developed countries were mainly preoccupied with the negative impacts of industrialization, while developing countries viewed the North's environmentalism as a blatant threat to their development objectives. They explicitly proclaimed the right to economic and social development and said that environmental concerns could not be used to limit this right."[33] In other words, everyone has an equal right to exploit nature, no matter the consequences. Settler colonialism has no national limits.

It is impossible to overestimate the ideological significance large dams gradually assumed in the twentieth century. Large dams "epitomized the conquest of nature by technology."[34] Repeatedly, this conquest assumed a national coloring that trumped the peculiarities of individual political systems: "The control of a great river through a mega-dam is one of the most emphatic advertisements of collective human prowess, national clout and the muscle of the central state."[35] Dams are "political symbols: symbols of the conquest of nature, of progress, and of the modern state. And, especially, symbols of national empowerment and achievement."[36] Dams could be and were cited as examples of the superiority of any and every type of political system or regime. Once again, Lilienthal's title is a case in point. In the former Soviet Union, Josephson observes, the first massive hydroelectric project, Dneprostroi, "became a symbol of what centralized economic planning and political will could accomplish"; in the United States, before Lilienthal so baldly stated their broader significance, dams were initially "symbols of the New Deal rebuilding of America."[37] In the mid-twentieth century in newly Communist China, writes Simon Winchester, "the dam was seen by Mao and his allies as perfect propaganda for the promotion of his authority and power."[38] In revolutionized Egypt in the 1950s, explains John Waterbury, "the new regime [of Gamal Abdel Nasser] sought a spectacular gesture to signal its visions and intentions to the Egyptian people and to the world."[39] The

Aswan High Dam project was precisely the kind of gesture desired. Dams represented political and economic success, whatever the system of government that had produced them; literature stood ready to assist in fostering such ideas, bringing a human coloring to technological achievement.

In retrospect, it seems inevitable that large dams, in their massive apparent simplicity, began to evoke comparisons with the Great Pyramids of Egypt or at the very least with large temples—often to the detriment of the pyramids and temples. Grand Coulee Dam on the Columbia River gained fame in part because its volume was the first to exceed that of the Great Pyramid of Cheops.[40] In Egypt, a comparison between dams and pyramids was de rigueur. The British magazine *The Spectator* declared that the Aswan Low Dam, built at the very beginning of the twentieth century, was "as great a memorial as Cheops' Pyramid, and of considerably more use to Egyptians."[41] In the mid-twentieth century, Nasser was said to have frequently observed: "In antiquity, we built pyramids for the dead. Now we will build new pyramids for the living."[42] In other words, dams were obviously superior to pyramids; they were functional, not memorial. At about the same time, in India, Jawaharlal Nehru liked to describe dams as the "temples of modern India."[43] Some thought dams resembled Buddhist stupas.[44] In France, the Donzère-Mondragon Dam, completed in 1952, evoked comparisons with great cathedrals. Such comparisons were still being made decades later. In his study of rivers, dams, and the conservation movement, Tim Palmer observes that "tourists stand awestruck at the face of [Hoover] dam as at an Egyptian pyramid."[45] Dams were "America's cathedrals, its castles, its pyramids," declares Stephen Grace in his study of the role water played in the American West.[46] In another work on water in the American West, Marc Reisner speculates that "when archaeologists from some other planet sift through the bleached bones of our civilization, they may well conclude that our temples were dams."[47] Dams played a role in both colonization and postcolonialism, symbolically and in actuality, a role that continues to exert a profound influence.

While simple embankment dams had been constructed for thousands of years, by the twentieth century there were multiple possibilities—some novel—for containing rivers and various reasons to do so. In many places, the desire for flood control or improvement of navigation was paramount. In others, the facilitation of irrigation and creation of reservoirs was most important. Use of dams for power, especially hydroelectric power, was a later

14   INTRODUCTION

but critical development. Initially, there were embankment dams of earth, rock, or a combination of the two. Later came solid masonry gravity dams, like Hoover and Grand Coulee, made of immense quantities of masonry or concrete and situated in the most appropriate geological locations—provided, very important, that their construction did not threaten the existing achievements of the affluent. No dams were to be built in places inhabited by wealthy landowners. There were also hollow masonry gravity dams, timber dams, steel dams, and arch dams. Hollow masonry dams required more elaborate engineering design; arch dams, like the Swiss Mauvoisin Dam, can only be built in narrow canyons and demand the greatest architectural sophistication.[48] Engineering enthusiasm accompanied these important technological shifts in a wide variety of political arenas.

Earlier dams were generally built with a single purpose in mind, like flood control. In the twentieth century, multi-use dams that could, for example, simultaneously improve navigation, assist with irrigation, provide flood control, and serve to generate hydroelectric power became popular. There was no limit, it appeared, to what mega-dams could do. Ultimately, the notion of river basin planning, of the building of multiple, sometimes dozens of dams, became highly influential. River basin planning involves the construction of a series of dams along a river and possibly its tributaries; it pursues not only dam building but the industrial and agricultural development of the areas adjoining the river. The TVA became a worldwide poster child for river basin planning; as such, it exerted tremendous influence in the second half of the twentieth century and evoked the jealousy of, among others, the Soviet Union. The TVA provided a development model that could supposedly be adapted to any geographic circumstances and often was, albeit usually in a rushed, ill-conceived, and uncritical fashion. All that was necessary for success, supposedly, was funding and the national will to engage in technological development for the supposed good of a nation's inhabitants. As Lilienthal proudly declared about the TVA, "For the first time in the history of the nation, the resources of the nation, the resources of a river were not only to be 'envisioned in their entirety'; they were to be developed *in that unity with which nature herself regards her resources*—the waters, the land, and the forests together."[49] Nature was thus hijacked as a partner to modernity, as a patron of development. Such enthusiasm on the part of Lilienthal and others inspired the creation of literary works devoted to the primacy of development and the subsidiary role played by nature, works that captivated the attention of a variety of readers and helped convince them of the virtues of modernity.

The TVA was a federally authorized project launched in 1933 in connection with the much-touted New Deal. Even at that point, before the Cold War developed, fears of socialism helped inspire developmental aspirations. The Tennessee Valley was an economically disadvantaged and depressed region even before the Great Depression. The project was designed to improve navigation, assure flood control, generate electricity, and foster economic development, both agricultural and industrial. As a regional planning agency, the TVA encompassed most of Tennessee and parts of Alabama, Mississippi, Kentucky, Georgia, North Carolina, and Virginia. It was the first federal project of its kind and was highly controversial because of disputes over the pros and cons of government ownership of the utilities involved. This was denounced by naysayers in some quarters as a form of incipient socialism, but they were quickly silenced. Others—many writers among them—convinced that private utility companies were unregulated and rapacious, welcomed the move because of its apparent potential for changing society. A new and improved society, an implicitly model democratic one, was a desirable goal.

More than a dozen dams were constructed during the first decade of the TVA's existence. By the end of World War II, the agency had become the largest supplier of electricity in the United States. As "the first modern, large-scale effort in the world to plan and finance integrated regional development," the TVA quickly gained fame in the postwar years and played an important role in Cold War politics.[50] Visitors came from many countries to admire the awe-inspiring achievements of the democratic US government. Lilienthal founded an engineering and consulting firm that worked with clients from Iran to Nigeria to India in developing TVA-type projects. The concept of river basin planning was adopted with enthusiasm throughout what was then termed the Third World and became an important part of the US Cold War arsenal, a way of showing that capitalism could empower any nation's growth. What has been called "hydroelectric envy" dominated world politics,[51] much like "missile envy" in regard to nuclear weapons. Dams, dams, dams—the more, the better.

It was not only the United States that led the way in the worldwide hydropower competition. The former Soviet Union had been an equally significant player ever since the construction of Dneprostroi in the 1930s; in the postwar

16   INTRODUCTION

years, as the Cold War gained greater traction, the Soviets too began to engage aggressively in river basin planning—in part in competition with the United States—initially on the rivers of central Russia like the Volga and soon on the massive rivers of Siberia. Technology, it quickly became apparent, was a great leveler of political beliefs. Just as an ideological commitment to the domination of nature knew no political or national boundaries, so did river basin planning lend itself to adoption by a variety of political perspectives. As Josephson points out, Soviet hydroelectric power stations could reveal "the advantages of the Soviet system over those of capitalist countries and symbolized the qualitative difference between peaceful Soviet electricity and imperialist, militaristic capitalist energy."[52] No one political system had a lock on modernization. As an added plus, the Soviet projects were often gleefully, if carelessly, accomplished in much less time than similar projects had been in the United States. Those who wanted dams in other countries took notice. Later India, a would-be example of high modernism, provided a postcolonial model to emulate. The Hindi Indian feature film *Mother India* (1957), which glorifies big-dam construction, was widely viewed across the world for decades, making India's dams "a symbol for hope and progress across postcolonial Asia and Africa."[53] What India could do, an entire range of countries of diverse political persuasions could implicitly accomplish, following in US or Soviet footsteps—not as victims of colonialism but as political agents in their own right. Everyone could assume a colonialist role, either internally or externally.

The 1930s have been described as the "go-go years," the "glory days" of dam building.[54] After this auspicious beginning, dams' heyday lasted several decades. In *Silenced Rivers: The Ecology and Politics of Large Dams* (2007), the environmentalist and writer Patrick McCully analyzes in detail data collected by the International Commission on Large Dams (ICOLD). The numbers are staggering. There are more than 40,000 large dams (dams with a height of 15 meters or more) in the world. The 1960s witnessed the completion of an average of 1 such dam per day.[55] There are more than 300 major dams (defined as having a height of 150 meters or more, a volume of at least 15 million cubic meters, reservoir storage of at least 25 cubic kilometers, or electrical generation capacity of at least 1,000 megawatts). Tellingly, McCully points out that a volume of 15 million cubic meters is six times that of the Great Pyramid of Cheops.[56] Such gargantuan structures can be found on every continent but Antarctica and in dozens of countries; they surpass the pyramids because of their obvious usefulness, not just their size, and implicitly because they reflect a dramatic change for the better in society.

Inevitably perhaps, such large-scale and expensive undertakings as megadams attract political maneuvering, spawn large bureaucracies, and foment corruption in their wake. Bribery, pork barreling, price fixing, reckless cost cutting—big-dam projects invite it all and, unfortunately, to a great degree. The political prestige associated with large dams encourages government officials to underestimate expenses and downplay potential problems. The money involved may fuel corruption and subvert sensible business practices. As the twentieth century wore on and the number of geologically ideal, usable dam sites gradually diminished, such problems became more acute and the costs increased. Sadly, the very fact that the negative consequences of dams were becoming increasingly obvious paradoxically also fostered a readiness to gloss over or even conceal difficulties that might obstruct lucrative and prestigious construction plans. Here too some writers were eager to echo political goals in their work, to serve as cheerleaders for new technological developments.

In the middle of the excitement associated with the building of big dams, the uncomfortable truth that such construction often meant relocation of and tremendous financial harm to people whose farms and villages would be inundated by rising reservoir waters was initially frequently downplayed or ignored. Dams have not typically been built in areas inhabited by wealthy and politically influential elites; quite the contrary, in fact. There is a parallel here with the construction of superhighways in US cities. The usual twentieth-century dam site had a poor rural population, whose impoverished members, in turn, represented an ethnic or social minority. For the TVA, this meant impoverished Appalachians, who were indeed in dire need of economic assistance. In the American West, a more common pattern emerged. Native Americans with little political agency—such the Yakama in the Pacific Northwest and the Mandan, Hidatsa, and Arikara in North Dakota—often inhabited lands where dams were built. In Canada, members of First Nations, like the Cree in the James Bay area between Ontario and Quebec, had to move because of dam construction. Indigenous peoples were the ones generally affected by projects in Brazilian Amazonia and Soviet Siberia. In India, it was peasants from Indigenous populations referred

18   INTRODUCTION

to by the Indian government as "scheduled tribes," also known as Adivasi, or the Dalits, formerly known as "untouchables" in the Indian caste system. In Egypt, it was the Nubians, tellingly long regarded by many Egyptians with barely concealed contempt and overt racism. The ethnicity of those who need to move because of dam construction is rarely the same as that of those who will benefit from a hydroelectric project—a phenomenon typical of settler colonialism, which favors the colonizers and gives little thought to Indigenous peoples already on the scene but regarded as dispensable.[57]

Relocation or resettlement because of dam construction has multiple implications and consequences. Most obvious is the loss of one's home, which in the case of rural populations often also means loss of their source of livelihood, profound cultural upheaval, and psychological distress—particularly for the elderly and for women, who are often dismissed as collateral damage in the march of progress. When dams are built, communities may be fragmented, treasured cultural sites inundated. Graves need to be dug up and bodies transferred or cemeteries may be flooded, causing family members unspeakable trauma. When compensation is paid for lost land and homes, it may be far from adequate or appropriate and be provided only to those whom dam construction most obviously and directly affects. For example, in calculations of compensation, the incomes of people living downstream from a dam site, whose losses are less visible but whose livelihoods suffer as a consequence, are generally not considered. The number of people who have to move may be underestimated, in some instances deliberately, to facilitate implementation of a project. The process of resettlement and allocation of new land and homes may be poorly organized and subject to corruption. Little advance warning of the need to move may be given, leading to economically damaging chaos. Relocation sites may be of inferior agricultural quality, and possibilities for engaging in crucial economic activities like fishing disappear. New homes may be not just different from but inferior to the old. Many of those who are relocated, unable to sustain themselves economically in their new place of residence, may become migrant laborers or drift into urban slums.[58] In the second half of the twentieth century, such developments began to provide particularly fertile ground for literary diatribes whose intensity matched that of earlier panegyrics to modernization. In such contexts, supposed opportunities for the bereft often became increasingly difficult to identify.

For much of the twentieth century, the issue of relocation and its consequences for those directly affected received little attention. As Katrine Barber expresses in a study of relocation linked to hydroelectric projects on the Columbia River, it was generally "a backburner issue."[59] On the Columbia, the destruction by dams of salmon runs that century-long treaties with Native American peoples supposedly protected was widely ignored, as was the need to move some reservations.[60] Professionals with appropriate expertise involving human impacts were long uninvolved or saw their importance downplayed in planning for relocation: "Dam design is dominated by hydrology, engineering, geology and economics. Disciplines like sociology or anthropology or development studies are rarely given a professional role to play, and if they are it is too often a token one."[61] Until the 1990s, the World Bank, which is heavily involved in funding large dam projects, only occasionally enlisted the services of resettlement experts in appraising plans for mega-dams; after all, controlling nature was more important than coddling people. At a hearing held in 1989 by the United States Congress Human Rights Caucus, the Bank could not provide a single example of successful rehabilitation of relocated populations.[62] The demands of other priorities seemed far greater. In his environmental history of the twentieth-century world, J. R. McNeill suggests that "their political utility helps explain why so many uneconomic and ecologically dubious dams exist."[63] In his study of dams, Fred Pearce concludes: "It is the rural poor, along with their environment and natural resources of remote regions, that suffer in order for a few to benefit. By their nature, large dams and hydroelectric projects are amongst the least likely of 'development' initiatives to generate improvements in the lives of the rural poor."[64] The rural poor, it was long thought, should make sacrifices in the name of progress, even though they were unlikely to partake of that progress. The Australian journalist and Egyptologist Leslie Greener, writing in the 1960s about the efforts to salvage archaeological sites due to be inundated by the waters of the Aswan High Dam, commented bitterly: "Yesterday, little was known of Nubia, even by the archaeologists. Today there is world-wide interest in Nubia. But it too is largely theatrical: the dramatic threat of drowning; last-minute United Nations effort at rescue; the spectacular attempt to elevate spectacular Abu Simbel [Egyptian temples from the thirteenth century BCE]. It's enough to make you forget that people lived in Nubia too."[65]

20   INTRODUCTION

Greener's sarcasm is both obvious and appropriate. In the context of neo-colonial thinking, Indigenous inhabitants have little value.

The economic, social, and psychological upheaval caused by relocation was often rationalized as the acceptable suffering of the few for the benefit of the many, a long-standing argument used by exploiters from a wide range of political systems to justify the sacrifice of ethnic others. This argument was sometimes presented directly and unabashedly to those being resettled. In 1948 Nehru informed villagers affected by a dam project, "If you are to suffer, you should suffer in the interest of the country."[66] Two years later an Indian official, writing to a government minister about local objections to a project, explained that he had informed the local population that "they should not stand in the way of the construction of the project but should consider it a great sacrifice on their part, since by the sufferings, if at all, of a small number the country is going to prosper."[67] This self-serving logic, which was touted in many nations, must have sounded very hollow to listeners and impeded their efforts to practice agency in a new world.

The numbers of people affected worldwide by dam construction are truly staggering, if inexact. McCully declares, "Although the dam builders have not bothered to keep count, the number of people flooded off their lands by dams is certainly in the tens of millions—30 million would be a conservative estimate, 60 million more likely."[68] A leading expert on dams and relocation, the anthropologist Thayer Scudder, writes, "According to WCD [the World Commission on Dams] the number of those resettled in connection with large dams exceeds 40 million and may be double that number."[69] Even more explicitly than McCully, Scudder insightfully points out that the very absence of accurate numbers is itself indicative of a gross lack of institutional and governmental concern.[70] India and China provide particularly egregious examples of mass relocation. As of 1999, China had more than 10 million officially classified "reservoir relocatees."[71] McNeill suggests that in India alone, "dams and reservoirs displaced perhaps 20 million people between 1947 and 1992."[72] More recently, the figure of 40 million for India has been suggested.[73] Sadly, projects involving displacement of populations often had worldwide political and bureaucratic support; as dam building increased, so did the population impact. In his study of the World Bank, Bruce Rich—an American writer and lawyer who received an award from the United Nations for his exposure of financial development–masked greed—says that between 1978 and 1990, in India alone, the Bank funded projects that required the forced displacement of more than 600,000 rural poor.[74] Writers everywhere noticed such displacement and its concomitant alienation and began to write about it, often in highly moving terms that humanized the costs of displacement.

The social impacts of large dams cannot be easily separated from the environmental impacts, which in many instances are unanticipated and only gradually emerge. Most immediately obvious is deforestation and ensuing erosion and the loss of fertile bottomland, which, in turn, may have severe economic consequences. Deforestation may involve not only the reservoir site itself but adjacent land that is newly cleared by displaced inhabitants. In recent years, there has also been increasing recognition of the significant quantity of greenhouse gases emitted by reservoirs because of massive decomposition of rotting vegetation that was not removed from the reservoir site, grew up later, or was washed into the reservoir from upstream.[75] Despite the theory that large hydroelectric dams offer the cleanest energy, particularly in tropical areas, the emissions can be massive. Twenty years ago, Canadian researchers estimated that "reservoir emissions contribute 7 per cent of the total global warming impact of other known human-related releases of carbon dioxide and methane."[76] As the water quality in reservoirs declines, algae and water weeds may thrive. Among the most seriously undesirable of these are water hyacinths, which in tropical areas can rapidly infest reservoirs, stifling productive animal growth.

The creation of huge reservoirs also has a dire impact on species diversity, resulting in the disappearance of many kinds of fish, plants, and animals. The presence of a dam fragments ecosystems and interferes with breeding and other cycles, often to the point of destruction. One of the best-known examples in the United States is the detrimental effect dams on the Columbia River have had on the anadromous migratory salmon, which are born in freshwater, live much of their lives at sea, and return upstream to their birthplace to spawn. The ambitious and overly confident creation of fish ladders on dams, investment in hatcheries, and physical transportation of fish around dams has had limited success, contributing to a significant loss of income for fisheries and those who make their living fishing—many of whom are Indigenous victims of minimally honored or simply ignored nineteenth-century treaties made with settler colonialists. Plants and animals that grow and live beside rivers also suffer from the creation of large reservoirs and unseasonable discharges of water from them and are forced into ever greater competition with one another. All these subjects became rich topics for literary writings, for bringing environmental destruction to the forefront of readers' imaginations.

The problems relating to dams became clearer and more pronounced over time. Increased salinity of land and water is linked to dams. Large reservoirs may experience massive evaporation, which leads to greater salinity—a danger for agriculture and potability of water alike. The process of irrigation itself often promotes salinity of the soil, and the draining of irrigation water into reservoirs and rivers results in even more salinity. This is particularly a problem in hot, dry areas. The fact that the same water may be used repeatedly for irrigation creates a vicious cycle that leads to rapidly declining yields, especially from land near the mouth of a dammed river, where salt accumulates most dramatically.

Upstream especially, the increase in shallow and still water associated with dams has had an impact worldwide on the prevalence of certain diseases, such as schistosomiasis and malaria. The snails and mosquitoes that serve as vectors for these diseases thrive in warm, stagnant water. As with efforts to address the decline in fish populations associated with dam construction, attempts to control schistosomiasis and malaria have had limited success. Many thousands of people die annually from these two diseases.[77] This too has attracted the attention of writers.

It is also important to recognize that despite their impressive appearance, dams do not last forever, and their planners often overestimate how long they will function effectively. Siltation and sedimentation gradually undermine a dam's efficiency. Siltation refers to the sand, clay, or other types of soil found in running water. Excess siltation, which can be exacerbated by deforestation and erosion, is a form of pollution. Behind a dam, siltation leads to sedimentation. As the sediment increases, the storage capacity of the reservoir decreases. This process occurs inevitably with all dams; it is simply a question of how quickly it happens. Sediment removal is costly and complicated and thus rarely occurs. As a result, dams generally have a finite life span, which is often fifty years or less; dams are not pyramids in more ways than one. In addition, dams may be threatened by earthquakes, particularly in the case of very deep reservoirs, because of stress due to the weight of the water or unanticipated geological problems. A number of literary works have addressed such problems.

Downstream from dams, other problems arise. Perhaps to state the obvious, the fertile silt that backs up behind dams is lost to the river below a dam. This may lead to massive erosion, to what is termed the scouring of a riverbed. This, in turn, has a negative impact on downstream ecosystems; because of the increased flow of the river, ironically, this process worsens, not lessens, episodes of flooding. Scouring occurs surprisingly rapidly; riverbeds may become several meters deeper within a decade of closing a dam.[78] At a river's mouth, a delta may gradually shrink, and there may be further erosion along the nearby coastline. This in itself may have a range of undesirable and unexpected consequences. The erosion of the Nile delta, for example, has contributed to a tremendous decline in the sardine catch in the eastern Mediterranean.

Given the many negative impacts associated with dams, it is not surprising that resistance and outright protests—some literary—have increased in the past few decades. As McCully says, "The days when dams were seen as so obviously of great benefit to humanity that anyone who questioned them was immediately branded a communist / counterrevolutionary / antinationalist / outside agitator / deluded romantic / foreign spy are gradually dying."[79] Increasing protests have focused on both the social and environmental costs of dams. Such protests have sometimes emerged most powerfully in literary form.

In Western Europe and the United States, widespread objections to dams beginning in the 1950s came primarily from conservationists concerned with aesthetics, with the preservation of scenic wilderness areas.[80] A crucial effort was the successful lobbying led in the 1950s by David Ross Brower, executive director of the Sierra Club, against the construction of Echo Park Dam on the Colorado tributary Green River, which would have inundated Dinosaur National Monument. But this success was undercut by the almost simultaneous failure to prevent the construction of another dam on the Colorado River, Glen Canyon, about which Brower wrote: "Glen Canyon died in 1963 and I was partly responsible for its needless death. So were you. Neither you nor I, nor anyone else, knew it well enough to insist that at all costs it should endure. When we began to find out it was too late."[81] Finding out too late continues to be an operative concept where dams are involved.

Elsewhere in the world, the social impacts of dam construction were initially a greater cause for opposition than were the purely environmental

24 INTRODUCTION

impacts. In India, for example, peasants forced to relocate had been resisting dams since the 1920s, but generally with little effect or visibility, given their lack of political clout.[82] By the 1980s and 1990s, however, popular resistance had become more vocal and in some instances more effective. An important milestone in India in the 1980s was the creation of the coalition of environmentalists, students, and local peoples called the Narmada Bachao.[83] The Narmada River Project was a massively extensive plan for river basin development. The protests, led by the social activist Medha Patkar and at times involving 60,000 people, eventually led to the withdrawal of financial support by the World Bank—a tremendous path-breaking symbolic shift in priorities. In China, in contrast, in 2000 a protest by 1,000 relocated peasants over corruption involving the Three Gorges Dam project resulted in violence and military intervention.[84] Obsessed with development, the Chinese establishment has yet to fully acknowledge the potential human and environmental harm that may result from dams. Both the Narmada and Three Gorges projects continued, but unquestioning support for big dams has increasingly waned in recent decades, and the social and ecological concerns have often coalesced and gained political attention—although by no means everywhere.

Tellingly, even some of those who had previously waxed enthusiastic about large dams began to have doubts. The enthusiasm for mega-dams expressed in the United States and the Soviet Union in the 1930s and 1940s began to wane, beginning in the United States. Elsewhere, as early as the end of the 1950s, Nehru surprisingly expressed fears that "we are suffering from what we may call 'disease of gigantism'" and argued for small irrigation projects and hydroelectric plants.[85] Gone were the invidious comparisons with the pyramids. Many writers were among those whose opinions of mega-dams began to shift in a negative direction. In a new preface to *The Colorado* written in 1984, the American author and devotee of the Southwest Frank Waters writes about Hoover Dam: "A mammoth technological marvel, it evoked my own extravagant praise and admiration with that of the entire country. Since then I have come to regard it as the first of our misguided attempts to dominate the entire natural world of the river."[86] Perhaps the most succinct criticism comes from the poet, scientist, and activist Kenneth E. Boulding: "The more we dam the rivers, then the sooner we are damned."[87]

The striving to dominate nature may not be the best approach to life on earth adopted by human beings. Literary interest in dams has been great and widespread, from the glory days until the present. In the United States, the rivers whose dams have generated the most literary interest are the Colorado, the Columbia, the Missouri, and the rivers of the Tennessee Valley. In the former Soviet Union, the dams of the Ukrainian Dnieper, the Volga, and the Siberian Angara have been written about repeatedly. In Africa, the Aswan Dam and its impact have evoked responses for more than half a century, in part because of fears of terrorism. In China, the dams of the Yangtze, especially the Three Gorges Dam, often occupy pride of place. Dams on various Indian rivers have been featured in novels, songs, and films—at times in ugly and contemptible ways.

This study examines numerous literary works published over the past century that focus on dams and their social and environmental impact. Most of the novelists, poets, and essayists who have been responsible for these works are North American and Russian. As a Slavicist, I can read what has been published in a handful of European languages. Some of the exciting literature about dams that Egyptian writers of Nubian descent, as well as Chinese and Indian writers, have produced has fortunately been accessible to me in translation. Such works, at times written by representatives of those most negatively affected, are critical to understand the worldwide literary response to the growing impact of large dams—numerically, socially, and environmentally.

Chapter 1 has provided background information needed to understand the growth in the desire for mega-dams worldwide and has stressed literature's unique ability to promote such understanding. I have discussed attitudes expressed toward nature over time, in particular the insidious development of high modernism and the decision made to attempt to control nature, of which large dams are a prime example. I have also discussed in general terms the relocation of peoples made necessary by dam construction, often a primary consequence of settler colonialism; the negative environmental impacts of large dams, many of which were initially unknown or underestimated; and the role literature can play in drawing attention to and humanizing such problems. In the following chapters I will return to these topics in more detail and provide specific literary examples of thoughts expressed. Chapter 2 considers the predominantly North American, Russian, and Ukrainian works produced in the early and mid-twentieth century that endorsed dam-building projects in the United States and the former Soviet Union, often

26   INTRODUCTION

in the dramatic and grandiose, even pompous manner that became routine during the Cold War. In such works, dams assume a manifestly heroic aura. Chapter 3 is devoted to writings that focus on the trials and tribulations of the massive numbers of peoples dispossessed and displaced in the United States, Siberia, Egypt, India, and China—often as the victims of settler colonialism or its descendants. The largely detrimental social consequences of such dispossession and displacement warrant, and have gained, great literary attention. Chapter 4 examines various literary depictions of the environmental damage caused by dams, a huge and often underestimated problem that has social impacts as well. Chapter 5 looks at literary representations of acts of terrorism directed against large dams, including ecoterrorism. Ecoterrorism, or ecotage, to use the more neutral term, is perhaps a logical outcome of the harm produced by mega-dams in the twentieth century. Writers have paid attention to this development, with its vast, as yet unrealized potential for chaos and harm, on the one hand, and its dramatic way of drawing attention to environmental damage, on the other hand.

In recent years, a few dams have been successfully decommissioned. It would be simplistic to demand that all existing dams should be decommissioned and no new dams built. As many scholars and activists have argued, however, in the future dam construction needs to be approached critically, honestly, and with open eyes. My hope is that this study will assist in fostering an awareness of the complex history and complicated problems that have marked the impact of large dams worldwide and the future need for caution, care, and great thoughtfulness in considering the construction of more dams. Scientific environmental studies of dams are important, but many people do not read such works. Literary writings and their analysis in a broad, international cultural manner offer more accessible, personalized, and humanized views of the virtues and vices of mega-dams. A nuanced analysis of such works will help demonstrate dramatically what the general population is learning and can learn through literature about the construction of big dams and their profoundly poignant environmental, social, and human consequences.

2

The High Modernist Heyday of Mega-Dam Construction

Dneprostroi, Bratsk, Hoover, Grand Coulee, Tennessee Valley Authority—
there was a time, and for many Russians and Americans that time has not yet
ended, when the names of these dams and others like them evoked a heady
combination of delighted awe and national pride. The heyday of hydroelectric dam building that began in the former Soviet Union and the United States
in the 1920s gained tremendous world recognition for two very different and
antagonistic political systems, particularly during the Cold War. In the 1920s,
1930s, and 1940s and in the Soviet Union, and to some extent in the United
States, into the 1950s and 1960s, for the most part no one expressed doubts but
only confidence writ large that big dams could transform life and society in
positive and qualitative ways. In both countries, construction plans and efforts
generally met with an enthusiasm that at times bordered on the fanatical. In
both countries, journalists, essayists, novelists, and poets played a central role
in conveying the excitement and sense of achievement associated with hydroelectric development as a very visible component of industrialization and of
modernization in general. Authors as diverse in their political beliefs as Zane

---

https://doi.org/10.5876/9781646425976.c002

Grey, Woody Guthrie, Fedor Gladkov, and Evgenii Evtushenko, along with many other lesser-known writers, visited dams during and after their building and penned accolades to both the dams themselves and to those who participated in their design and construction. Such literary narratives bolstered a societal sense of promising economic and cultural development. Dams, it was thought, would not just benefit but actually transform society for the better and serve as a beacon of progress for other countries—the Soviet dams of a bounteous socialism, North American dams of a democratic capitalism. Everyone could thrive because of mega-dams. All that was necessary was to follow the example of the Soviets or the Americans.

The essays, novels, poems, and songs that were produced in the former Soviet Union and the United States at the peak of large dam construction reveal both consistent topical commonalities and intriguing thematic differences. An examination of these literary commonalities and differences contributes to an understanding of the atmosphere and ethos of high modernism, an ideology whose expectations regarding the unalloyed benefits of technology and economic development had broad support in both the former Soviet Union and the United States for several decades. Works like Gladkov's *Energiia* (Energy, 1932–1938), Evtushenko's *Bratskaia GES* (Bratsk Hydroelectric Station, 1965), Grey's *Boulder Dam* (1963), and Guthrie's "Roll on, Columbia" (1941), despite the radically different political views of these authors and others, capture the spirit and assumptions of a technology-obsessed era that helped define the ethos of the twentieth century. Such literature reflects the political and social preoccupations of two very different nations and captures the reader's attention in a compelling way.

From Dneprostroi to Bratsk

In Russia in the nineteenth and early twentieth centuries, many schemes for improving riverine transportation by building dams were considered. In what is now Ukraine, the notorious Dnieper River cataracts, which extended for more than sixty miles, greatly hampered ship traffic on much of the lower river for centuries and prohibited easy and much desired access to the Black Sea. The cataracts were a particular thorn in the side of Russian imperial economic ambitions but a thorn whose removal was difficult. Schemes for management and literal drowning of the cataracts by means of dams were discussed at the turn of the twentieth century, but nothing substantive happened, a lack of accomplishment the Soviets later blamed—not without some justification—on perceived flaws in the stagnant and cumbersome tsarist system and an unthinking prejudice in favor of traditional comforts and often backward mercantile interests of the aristocracy over navigational and other technological improvements. When the Bolsheviks came to power in the October Revolution of 1917, they began to plan almost immediately to develop water transportation and hydroelectric power on various Ukrainian and Russian waterways. As mentioned in the introduction to this volume, Vladimir Lenin and his compatriots were obsessed with the possibilities for economic and, they believed, the concomitant social growth electricity afforded. As early as 1920 the plan known as GOELRO, the Russian acronym for State Electrification of Russia, was initiated. GOELRO sought to facilitate both industrialization and massive social transformation through the benefits of electricity. Expectations ran high.

The early Soviet hydroelectric project that commanded the most attention was Dneprostroi, one of several huge dams eventually built on the Dnieper.[1] Dneprostroi flooded the dangerous cataracts that had caused problems for centuries, thus dramatically easing navigation and simultaneously enabling the production of immense electric power that permitted the growth of heavy industry, a key early Bolshevik goal. More than fifty settlements were flooded in part or in their entirety to allow for construction of Dneprostroi, but no official concerns were expressed for this loss; if there were private concerns, they remained silent. On completion, the dam was almost 762 meters long and 61 meters high. The largest dam in Europe at the time of its completion, Dneprostroi assumed a truly mythic status in the Soviet imagination. In 1926 Leon Trotsky asserted with great confidence: "The Dnieper runs its course through the wealthiest industrial lands, and it is wasting the prodigious weight of its pressure, playing over age-old rapids and waiting until we harness its stream, curb it with dams, and compel it to give lights to cities, to drive factories, and to enrich ploughland. We shall compel it."[2] Harness and compel; Trotsky's statement exemplifies the high modernist confidence that a supposedly passive, unconscious nature exists for human exploitation, that nature must be tamed and trained to submit to technological goals. Dneprostroi became and was long considered one of the wonders of the new Soviet world.

The Dneprostroi hydroelectric project drew on Western experience and, very directly, on Western—especially American—engineering experts for both its design and construction. The tremendous hostility that marked the Cold War had not yet begun, and the Soviets, despite their contempt for capitalism, were willing to take advantage of Western expertise; in the West, despite fears of socialism that had existed at least since the foundation of the Soviet Union and the first Red Scare after World War I, US instruction of foreign engineers and workers was still considered desirable. A key component of the ideology that underpinned the construction of Dneprostroi, however, was the Soviets' conviction that they had embraced a qualitatively different approach to labor, in which the workers revealed initiative and an admirable enthusiasm for self-sacrifice for the sake of the collective good. This translated into a willingness to work as quickly as possible with great fervor; individual desires and interests were subordinated to general needs. In the Soviet view, this difference afforded a contrast with capitalist exploitation of downtrodden and miserable workers. The official Soviet conclusion about the Dnieper hydroelectric project was that Soviet engineers and workers not only productively adapted Western European and US methods but surpassed them, in large part because of the existence of a dramatically different political and social consciousness. This was truly a brave new world that would expose the capitalistic and imperialistic failures of the West and quickly surpass the West in its accomplishments. The propaganda devoted to Dneprostroi bears out Daniel Klingensmith's contention, in his study of Indian dam building, that "the politics of dam-building . . . does not fully make sense unless it is seen as a larger project of moral development and national self-testing as well as a search for bureaucratic or political power."[3] In other words, dam building was intended to be literally life changing. Literary works made a point of demonstrating this—politically, socially, and at the personal human level.

DneproGES (the Dnieper Hydroelectric Station), or the Dneprostroi Dam, was built near Khortytsia Island—the largest island on the Dnieper—and the small town of Zaporizhzhia, today a major city. Construction of the dam began in 1927. During the five years the dam was under construction, dozens of members of the Soviet intelligentsia visited the site. They included the greatly admired Maksim Gorky (Gor'kii in transliterations; pseudonym of Aleksei Peshkov, 1868–1936), best known for his novel *Mat'* (Mother, 1906), one of the major precursors of socialist realism—the officially sponsored

The High Modernist Heyday of Mega-Dam Construction   31

and often wooden artistic style that dominated Soviet cultural life from 1932 to the late 1980s. Although Gorky did not always see eye to eye with individual Bolsheviks, he was a longtime supporter of many of their economic and political goals, which he regarded as essential for national development. The first Five-Year Plan (1928–32), launched to forward both industrialization and the collectivization of agriculture, was deemed so successful that it was pronounced completed after only four years. Dneprostroi was a signature component of this perceived success.

Gorky traveled extensively in the Soviet Union during the late 1920s and early 1930s. He recorded his thoughts about the inevitability of an emerging wonderful new world in a series of essays titled *Po Soiuzu Sovetov* (Around the Union of Soviets, 1929), one of which is devoted to Dneprostroi. For Gorky, the project represents an unquestionable and unquestioned human triumph over nature: "At Dneprostroi the will and reason of the working people are changing the figure and face of the earth."[4] In his infatuation with the concepts of rationality and intentionality, Gorky dismissively contrasts the Soviet adoption of clear goals with ostensibly (and implicitly nonexistent) divine haphazardness: "In antiquity, they say, God performed such mighty tricks. He was a poor builder, we have to redo everything in our own, new way."[5] Note the emphasis on a new way, implicitly socialist in character but imbued with the relentless drive of high modernism. Dismissive of any romanticization of nature, Gorky concludes his piece on Dneprostroi with a passionate and uncritical affirmation of the necessity for the domination of nature: "There is a poetry of 'a confluence of nature,' of immersion in its colors and lines, but this is the poetry of passive submission . . . There is also the poetry of the overcoming of the forces of nature by the force of man's will, the poetry of the enrichment of life through reason and imagination . . . This is poetry . . . for the sake of creators of new forms of social life."[6] What Gorky offers here is a kind of high modernist template for appreciating dam construction and the wonders of a new kind of labor—socialist in its essence—that would underpin Russian and East European literary treatment of dams on the Dnieper and other Soviet rivers for much of the twentieth century, becoming a hallmark of Cold War thinking and Soviet socialist realist literature.

Dneprostroi was the subject of many paintings, poems, and novels. Gorky's visit to the dam site was memorialized in a painting done in 1951 by Pyotr Ivanovich Kotov, *Maxim Gorky at the Construction of the Dnieper Hydroelectric*

Station, in which an obviously proud author stands on a bluff admiring the massive construction site and the river in the distance; the painting exemplifies socialist construction in all its glory (figure 2.1). Another famous early Soviet writer, the poet Vladimir Maiakovskii (1893–1930), had, like Trotsky, enthusiastically heralded the greatness of Dneprostroi in 1926—when the project was still in its planning stages—in the poem *"Dolg Ukraine"* (Duty to Ukraine, 1926). A committed, if at times critical, Bolshevik sympathizer, Maiakovskii had visited the United States in 1925 and produced a heartfelt encomium to the Brooklyn Bridge, a structure that in the poet's eyes symbolized the true beauty and significance of modernity. "Duty to Ukraine" was penned in much the same vein and offers the reader a Soviet equivalent of the Brooklyn Bridge. Gone are the days of the Dnieper's wild and purposeless Cossack past, thinks Maiakovskii. Instead: *"Dnepr / zastaviat / Na turbiny tech' / I Dnipro / po provolokam-usam / elektrichestvom / techet po korpusam"* (They will force the Dnieper to flow into turbines. And the Dnipro will flow through buildings with electricity along mustache-wires).[7] Nature will serve humanity in industrial glory. Maiakovskii's use of both the Russian *Dnepr* and the Ukrainian *Dnipro*, a common device in contemporary literary works about the dam, points to a utopian vision of the river as transcendent of narrow ethnicity—that is, a truly socialist and broadly Soviet creation. In the broadest sense, given Maiakovskii's admiration of the Brooklyn Bridge, the Dnepr/Dnipro contrast also points to the transcendence of nationalism in general, a key feature of high modernism.

Another dedicated socialist realist writer, the poet Aleksandr Bezymenskii (1898–1973), spent several decades refining the narrative poem *Tragediinaia noch'* (A Tragic Night, 1930–63), a lengthy account of how participation in the Dnieper project enables a large cast of characters not only to build a dam but to develop a novel and righteous political sense that guides their behavior. For example, even an American involved in the project eventually comes to see the communist light, while an old man who tours the construction site recognizes the benefits the Soviet government will bring to the people, and an engineer who had been stifled in his dreams for riverine development by tsarist backwardness and selfishness finds, to his delight, that his talents are appreciated by the new regime. To add a bit of drama to this potentially tedious panegyric, a few reactionary "bad guys" quit the scene when their malicious attempts to foil progress are thwarted. The inclusion

FIGURE 2.1. Pyotr Ivanovich Kotov, *Maxim Gorky at the Construction of the Dnieper Hydroelectric Station*, oil on canvas, 1951. Moscow, State Central Literary Museum, http://bit.ly/4720eov.

of such incidents, to add dramatic interest to otherwise tedious narratives, became a common feature of Soviet works devoted to the wonders of the new approach to industrialization and agricultural development.

Bezymenskii provides a brief account of the Dnieper region's history and a careful outline of the project, in which he gives very specific credit to the engineers involved. Engineers were already well on their way to being perceived as heroic figures, in a socialist context and elsewhere. The most glowing treatment is reserved for Aleksandr Vinter, the chief engineer at Dneprostroi, who presents the viewer with *"mysl'iu vdokhnovlennyi / Glubokii vzgliad . . . Sliian'e sily nepreklonnoj / S teplom rebiach'ei dobroty"* (a profound gaze inspired by thought . . . a mixture of unbending strength with the warmth of a child's goodness).[8] *"Neutomimyi rytsar' toka"* (an indefatigable knight of electric current), Vinter understands perfectly his responsibility to members of the working class.[9] As will be seen, laudatory attention to the role played by engineers also assumes great importance in American literary accounts of dam construction. Engineers were modern heroes—knowledgeable, focused, implicitly opposed to the crudity of nature, and supporters of an international technological utopia. Like Gorky, Bezymenskii makes sure to highlight the positive evolution of the work process itself, not just the

34   THE HIGH MODERNIST HEYDAY OF MEGA-DAM CONSTRUCTION

building of the dam: *"I Dnepr drugoi/ I liudi drugie/ I trud ne takoi"* (Even the Dnieper is different, people are different, and labor is not the same).[10] The backward differentiation of public and private has vanished in the face of a true communal spirit, the spirit that ostensibly embraced the new Soviet Union and was alien to exploitative Western capitalism.

Also like Gorky, Bezymenskii stresses the victory over nature Dneprostroi represented. In a description of an actual occasion in May 1931, when unusually heavy spring floods overtopped the dam while it was under construction, organized human opposition defeated the Dnieper's ignorant cunning. Bezymenskii uses the somewhat awkward but telling metaphor of an electric knight, an image with militaristic overtones, to describe the dam's ultimate victory over the unproductive landscape and nature's unthinking attempts to hinder development.

Another Soviet Russian writer who was profoundly dedicated to forwarding the industrialization of the Soviet Union was Fedor Gladkov (1883–1958). Gladkov's best-known novel is *Tsement* (Cement, 1925), which, like Gorky's *Mother*, was an exemplar of socialist realism *avant la lettre*. The title says it all—*Cement* is the heartwarming story of the restoration of a cement plant filled with characters who, like Bezymenskii's, successfully cultivate their socialist values in a variety of contexts. Gladkov spent years observing the construction efforts at Dneprostroi. He willingly made repeated changes to *Energy*, the novel that resulted from his observations, in an attempt to satisfy the shifting politically motivated objections of critics as the contours of Soviet socialism became more rigid.

The title *Energy* refers literally to the production of hydroelectric and industrial power. Figuratively, it evokes the psychological and physical energy the novel's characters dedicate to producing an entirely new and distinctly Soviet existence. Individual party members, administrators, engineers, and workers gain an appreciation of the potential afforded by collaborative socialist labor and are themselves thus able to make increasingly enlightened contributions to the Dneprostroi project. As in *A Tragic Night*, there are a few obligatory would-be evildoers whose efforts are doomed to failure.

Although the Dnieper itself makes surprisingly few explicit appearances in the several hundred pages that comprise *Energy*, when it does it conforms to a familiar image of hostile, even monstrous, nature. Early in the novel, one of the characters almost drowns while swimming in the river:

Suddenly he was thrown off to the side by the explosion of a whirlpool. The water began to seethe and sing. It began to play tricks with him, it grabbed him by the legs and arms, and whipped him in the face.

Suddenly he saw that the water was alive, incomprehensible, predatory, and quick sighted, and that its eyes were granular and opaline like those of a dragonfly. It was watching him from every quarter, all scaly and in convulsions.[11]

Nature red in tooth and claw indeed, monstrous and chaotic. Gladkov's imagery here verges on the bizarre, but its intent is unmistakable. Nature is an enemy par excellence, and the riverine landscape beside the Dnieper offers nothing remotely evocative of pastoral charm: "The brown hills, bare and clayey, the granite blocks wrenched from the depths, the river in its high stone shores—it all slumbered cheerlessly and inhospitably in primordial wildness. Faded peasants, with chapped faces just as brown as the hills and the granite, lived in ravines, hiding from winds and floods."[12] It seems as if the Stone Age has barely ended along the Dnieper. As in *A Tragic Night*, one of the few times the river is specifically mentioned in *Energy* is in a description of how it threatens construction of the dam. The river constitutes a hostile and irrational natural force, to which implicitly only human efforts, thinking, and planning can effect a positive change.

Another Soviet author who wrote with officially well-received rapture about Dneprostroi a few years later was the Ukrainian author Iakov Bash (pseudonym of Iakiv Bashmak, 1908–86). Bash worked at Dneprostroi from 1928 to 1932 as a carpenter and brigade leader and then embarked on a university education and a career as a writer. His best-known novel about the Dnieper hydroelectric project, *Hariachi pochuttia* (Warm Feelings, 1947), initially appeared in Ukrainian. Translated into Russian in the early 1950s, *Warm Feelings* went through multiple Russian and Ukrainian editions.

Warm Feelings frequently calls to mind *Energy*, but it contains many more detailed descriptions of both the river and the actual dam construction process than does Gladkov's novel. Like Gladkov, Bash underscores the tremendous difference between the useless imperial past and the productive Soviet present. The battle with the river is again cast in explicitly military terms. Constantly exploding sounds of dynamite lead to a characteristic comparison: "This persistent siege of the river, these explosions again recalled the front."[13] But now the river is the front, not a source of any kind of natural beauty. Once again the record-breaking spring flood of May 1931 is described, this time using overtly militaristic imagery: "The waves came like mountains. With a threatening rumble they rushed against the barriers, they flung themselves against the crib-work walls, and tried to break through into the foundation pit . . . The hydroelectric station resembled a huge, bustling port that was being rapidly evacuated in the face of a threatening attack."[14] Dangerous and frightening as it is, though, this anthropomorphized but essentially unthinking river cannot match intelligent, politically astute human efforts. When the dam is complete, the river falls obediently into line: "It quietly and submissively turned toward the sluice-gates and obediently, like a tamed giant, bent its neck into the pipe, flowed down, and turned the turbine faster and faster."[15] The giant even has a name—*Slavutich*. This is a traditional folk name for the Dnieper, a patronymic meaning "son of glory" (*slava*). Toward the end of *Warm Feelings* an old river pilot proudly asserts "*Slavutich* has been chained!"[16] Implicitly, this is what nature, which strives to be independent and intrepid, really deserves—to be imprisoned and enslaved by the superior thinking of socialist human beings.

Dneprostroi was, in fact, an astounding achievement. Tragically, less than a decade after its completion, the dam had to be deliberately decommissioned. The German invasion of the former Soviet Union at the beginning of World War II meant that Ukraine became a huge battlefield. In August 1941, when the Germans were on the verge of seizing the Dnieper hydroelectric dam, retreating Red Army forces—working together with administrators and engineers—dynamited the dam, unwilling to abandon the highly strategic structure to the enemy (figure 2.2). Some of the dam's internal components were shipped to the East. When the Germans were forced to retreat two years later, they spitefully vandalized the remains of the dam. It was, however, rebuilt and enlarged after the war and retained its status as a symbol of Soviet socialist success, of transcendent modernism.

The destruction and reconstruction of Dneprostroi endowed the dam with a distinct patriotic aura in the Soviet consciousness. One of the many postwar Soviet literary works that commemorated the battle for Ukraine was a novel by the Ukrainian writer Oles' Honchar (1918–95), *Liudyna i zbroia* (People and Arms), which was published in Russian as *Chelovek i oruzhie* (1960) before it appeared in Ukrainian. The work contains a vivid account of the intentional Soviet destruction of the hydroelectric station. Dneprostroi

FIGURE 2.2. Max Alpert, *Destroyed Dnieper Hydro Electric Power Station*, photograph, January 1, 1943. Wikimedia Commons: RIA Novosti Archive, http://visualrian.ru/ru/site/gallery/#604045.

is portrayed as the still beating heart of a tormented Ukraine. One of the soldiers involved sees the dam as an extraordinary realization of the river's potential, a harbinger of the Soviet wonderland still to come, despite the German invasion:

> The huge concrete crest of piers across the entire Dnieper, the cranes over the dam, and the hydroelectric station on the right bank that resembled a fairy tale palace . . . all this, together with the granite of the shores, the azure of the Dnieper, the green hills of Khortitsa Island, and the high dome of the sky, flowed together into a single whole, and stood like one harmonious

creation begun by nature and completed by man. Force and Harmony. Light and purity. It seemed that not a speck of dust had ever fallen on this structure, on everything that shone here with newness, with a kind of festiveness. It seemed that this sunny picture had been drawn from somewhere in the future as a model of what would someday triumph over the land.[17]

This is a portrait of high modernism at its most wondrous. In its glorious riverine setting, Dneprostroi here becomes a virtual temple and harbinger of high modernism whose destruction seems unimaginable to the Soviet soldiers. For the dam's engineers, the structure and its operations represent the pinnacle of their accomplishments, but they console themselves with the thought that their own assault on the dam's capacity will forestall German desecration. Nonetheless, when the destruction actually takes place, it is as if the unenlightened nature-dominated past gains a temporary victory over the socially progressive present: "Like the voice of past times, in the dusk over the Dnieper there resounded the threatening rumble of the rapids, a deep, seething roar, in which there was something primitive, wild, and gloomy."[18] Once again, ugly nature has emerged in its elemental and unproductive, if powerful, essence. The roar of the rapids provides a fitting accompaniment to the violence and destructiveness of the German invasion. Only the knowledge, from the perspective of the postwar period, that the Germans would be defeated and Dneprostroi rebuilt lessens the traumatic impact of this description.

World War II slowed and played havoc with Soviet hydroelectric ambitions in multiple ways. Several hundred miles to the east of the Dnieper, the Volga dominates the Russian landscape. At 3,692 kilometers in length, the Volga is the longest river in Europe, and it is the longest river in the world that does not flow into an ocean; it flows into the Caspian Sea. From the 1920s on, the Soviets sought to redefine the huge Volga basin as a markedly Soviet space. This involved industrialization, collectivization, and the development of hydroelectric power. Once again, Gorky captures the ideological imperative at issue in his observations on the Volga in "Around the Union of Soviets." For Gorky, the major goal of the working class conforms to a familiar agenda, analogous to that of Dneprostroi: the need "to transform the blind and tempestuous forces of nature into its rational servants."[19] The Volga is another Slavutich, in need of chains that human hands have

The High Modernist Heyday of Mega-Dam Construction   39

thoughtfully wrought. The late 1930s witnessed multiple efforts in this direction on the Volga, including the construction of hydroelectric dams on the river's upper portion. Significantly, the assistance of foreign engineers was no longer considered necessary or desirable.

Like the Dnieper, Russian engineers had eyed the Volga and considered its possible damming during the late tsarist period. As in the case of the Dnieper, establishment figures had often selfishly resisted such intentions.[20] The plan that was developed, under Bolshevik auspices, envisioned a cascade of dams erected from the upper northern reaches of the Volga all the way down the river. Dam construction and river linkage were to be accompanied by the construction of lengthy canals, like the Volga-Moscow and Volga-Don Canals. The opening of the latter in 1952 enabled ships to pass easily from one to any of the other European Russian seas and was indeed a huge accomplishment.[21] Soviet propaganda made much of this achievement.

In the 1930s hydroelectric dams began to be built along the northern Volga near the cities of Ivankovo, Uglich, and Rybinsk as part of what was termed the Big Volga Project, a plan for a series of large dams along the motherly river that would enable industrial growth and the dissemination of electric power over a vast area. At Rybinsk, which was put into operation in late 1941, after World War II had already started, a large statue of Mother Volga—an anthropomorphized representation of what Russians to this day regard as the quintessential Russian river—overlooks the reservoir and seemingly gives her blessing to hydroelectric power. As happened with the Dnieper, however, such developments came to an abrupt and violent halt with the outbreak of the war.

After World War II ended, the Big Volga Project quickly resumed. The terrible destruction suffered by the Soviet Union during the war, combined with the rapidly emerging Cold War, gave the project particular urgency. As the Cold War developed, the Soviets felt strongly that they needed to show as quickly as possible that they could match US technological development. In the Soviet Union as elsewhere in the postwar period, dams and hydroelectric power appeared to promise economic and social progress on a massive scale. A keen sense of competition with the United States was heightened in the Soviet Union in part because of the aggressive US marketing of the Tennessee Valley Authority (TVA) as a model for rural development and capitalist accomplishment. The fact that, unlike Dneprostroi, the gigantic dams that were built on the middle and lower Volga were exclusively domestic accomplishments was a great source of pride (no more need for foreign experts).[22] The reality that prison labor played a major and unsavory role in these projects was generally unmentioned. Only later did this ugly feature of mid-twentieth-century Soviet dam construction begin to receive attention in print.

As with Dneprostroi and other Soviet industrial collective farming accomplishments, Volga big-dam construction seemed to demand equally lengthy and at times tedious novels. In the course of the 1950s, Fedor Panferov (1896–1960), a Soviet Russian writer committed—like Gladkov and Bash—to propagandizing socialist achievements, produced the trilogy *Volga matushka-reka* (Little Mother Volga). In the 1930s Panferov had authored a glowing account, *Bruski* (1928–37). Like *Bruski*, Panferov's trilogy *Udar* (The Blow, 1953), *Razdum'e* (Meditation, 1958), and *Vo imia molodogo* (In the Name of the Young, 1960) concerns the challenges and ultimate rewards of the controversial Soviet collective farm initiative. The novels give space to the Big Volga Project and its positive impact on the expansion of agriculture through irrigation along the lower Volga. Throughout the narrative, the major protagonist, Akim Morev, is a Communist Party secretary, a typical heroic figure in Soviet construction novels—literary compositions that stand in direct opposition to tales of capitalist oppression of workers. Morev has dedicated his working life to encouraging harmonious and productive collectivized labor. He provides a true model of thoughtful and focused socialist commitment.

By the 1950s, Soviet hopes for hydroelectric power were even higher than they had been in the heady 1930s. Truly utopian dreams are expressed by the agronomist Ivan Bakharev, whom Morev meets and sails with down the Volga on his way to his new assignment near Astrakhan, a city on the southernmost portion of the Volga near the Caspian Sea. Predictably, for Bakharev, nature is "an angry enemy" that "conceals within itself evil forces, with which we will have to fight and fight."[23] Bakharev is certain that not only will canals, dams, and massive irrigation turn the dry steppes along the lower Volga into a tremendous agriculture wonderland but that soon there will be even more amazing developments. Much discussed in the 1950s and for many years thereafter—not only in the Soviet Union—were the glorious possibilities associated with reversing the course of Siberian rivers from north to south, thus providing limitless water for agriculture and industry in southern Siberia. Bakharev embraces such visions unquestioningly: "We

The High Modernist Heyday of Mega-Dam Construction   41

will change the courses of the rivers of Siberia from the North, we'll inundate the gigantic Baraba steppe [in western Siberia] . . . We'll connect all the rivers."[24] In Bakharev's eyes, there are apparently no limits to the transformations human beings can impose on the natural world. Once again, we witness high modernism on steroids.

For the collective farmers of *Little Mother Volga*, however, it is above all agricultural successes, rather than navigational improvements and hydroelectric power, that will represent the greatest advantages afforded by the Big Volga Project. In this context too, images of destruction and control of willful nature play an important role. Morev talks enthusiastically to the famers about the necessity to "destroy the desert's breath, in other words, to change the climate of the Volga region and transform little mother Volga into a real feeder of the country."[25] The Volga had long been popularly construed as the nourisher of the Russian people. Morev's suggestion implies that humanity can, unsurprisingly, improve upon nature, producing a true cornucopia.

Like the Dnieper, though, the Volga does not always conform to human intentions, which perhaps makes the supposed human victory over the river that much sweeter and at least more dramatic. At one point in the novel, an underground river tributary unexpectedly brings huge quantities of mud to the surface that threaten to stymie irrigation processes. The collective farm workers and others rush into action, and at the end of the trilogy thousands of enthusiasts attend the opening ceremonies for a new canal. "We have conquered," declares a triumphant Morev, in a manner reminiscent of his many socialist predecessors.[26] This time, though, it is Mother Volga, not Slavutich, who has been defeated.

The Big Volga Project was of immense importance both economically and culturally. By the late 1950s, though, it was dwarfed in both the official and popular Soviet imaginations by what was happening in Siberia. In their truly impressive immensity and hydroelectric potential, the many rivers of Siberia were—like the Dnieper, Volga, and other western Russian rivers—the objects of engineering interest well before the revolution. The river that attracted the most attention was the east Siberian Angara, which is the only river that flows out of Lake Baikal, the world's deepest lake. This means that in contrast to most rivers worldwide, which exhibit a riverine origin pattern of gradual buildup from small streams and spring ice melt, the flow rate of the Angara is both large and unusually consistent—a natural gift practically begging for exploitation. For the Angara, there would be no need to construct a large artificial reservoir to feed its hydroelectric stations. The Baikal, which contains more than 20 percent of the world's unfrozen freshwater and is a little more than a kilometer and a half deep, was already there with water aplenty. Soviet hydrologists called Baikal and the Angara "a Siberian marvel, the pearl of Soviet hydroenergy."[27] Such grandiose assessments were not limited to Soviet commentators. In 1969 an American author asserted: "The Angara is probably the most ideal river for hydroelectrical development anywhere in the world."[28] Strong words indeed. The Cold War did not always prevent admiration of the enemy's resources and accomplishments, particularly in a high modernist context.

Even before the revolution, engineers recognized that the Angara would be key to the development of eastern Siberia, to its transformation into a socialist paradise. In the early 1920s, intensive analysis of the river's potential began.[29] By the early 1930s, plans for the construction of a cascade of dams along the Angara—the first near Irkutsk, the city closest to Lake Baikal—were discussed. As with the Big Volga Project, World War II prevented the immediate implementation of such plans. At the same time, though, the transfer of many industries and people to Siberia during the war gave even greater impetus to the goal of exploiting the Angara. Construction of the first dam in the planned cascade of dams, only sixty-four kilometers from Lake Baikal, began in 1950. Construction workers who had helped build dams far to the west in Russia flocked east to the Angara, as did many young Soviet citizens for whom this was their first experience of the excitement of large dam construction—an excitement literature frequently underscores. By 1958 the Irkutsk or Baikal Hydroelectric Station was in full operation, producing a tremendous quantity of energy.

The irrepressible Gorky, of course, had positive things to say as well about the industrial promise the Angara afforded: "Capturing the imagination with their grandeur, fairytale pictures unfold of a future Siberia that the elemental force of the Angara will create when it has been made meek and mastered by the laboring energy of the people."[30] Once again, nature would succumb to human efforts, efforts implicitly inspired by socialism. The Angara would follow in the footsteps of the Dnieper and the Volga.

In 1957 the journalist Frants Taurin (1911–95) published the novel *Angara*. Taurin had served as editor of the Irkutsk Dam construction site newspaper,

The Fires of Communism. Angara systematically observes the norms of Soviet production novels like those concerning the Dnieper and the Volga. Painfully long and replete with the kinds of technical details Taurin absorbed as editor of *The Fires of Communism*, the novel has a huge cast of characters. Some are mature communists from the beginning; others gain greater political consciousness as the novel proceeds, and there are a few unsalvageable villains. There are many love interests, a few tragic losses of life, and some satisfying exposures of poor behavior. All this helps humanize the book's action. It is, however, the tremendous engineering achievement exemplified by the dam that occupies pride of place in the story of the building of the Irkutsk Dam.

In keeping with its unusual geological profile, that is, its emergence from Lake Baikal, the Angara is defined as qualitatively different from the beginning of the novel. While there are many rivers, each unique in its own way, the Angara stands out:

The pearl of Siberia—the Angara—is different from all rivers.

Having broken out of the stone ring of the Baikal mountains, it is born powerful right away; it combines the high water of a river of the plains with the swiftness of a mountain stream.

The proud force of the Angara is in keeping with a gigantic region. No wonder the Siberians love it so much. The Angara is like the soul of a Siberian—indomitable and brave, severe and pure.

Every river has its fate and its time. And now the time of the Angara has arrived.[31]

Everyone involved in the construction of the dam appreciates the formidability of the Angara, but they are also very sensitive—often in the unironic and pompous manner that especially marks later socialist realist novels—to their own importance as Soviet citizens and the historical value of their new type of labor. Their bureaucratic superiors at every level are likewise filled with a sense of self-importance. The head of the central institution in charge of dams opines: "The Angara is flowing into the future! The energy of the Angara is the foundation, the backbone of the new, most powerful industrial base of the Soviet state. Within a few years East Siberia will rank with the Donbass, the Urals, and the Kuzbass, and we, the hydro-builders, will be pioneers in the mastery of its innumerable natural riches."[32] Like countless officials in Soviet construction novels, this bureaucrat realizes that the dam project is not just an engineering project intended to dominate nature but a transformative socialist program as well, an important component of the Cold War conflict and a model for other nations. In describing to a subordinate how essential proper living conditions are for workers on the construction site, he insists "you are not building the pyramid of Cheops, but a hydroelectric station in the Soviet Union."[33] By implication, Soviet laborers are not Egyptian slaves, and, of course, a hydroelectric structure is superior to any useless pyramid. Implicitly, Soviet workers, because of their unique socialist work ethic, are superior to American workers as well.

Throughout *Angara*, the linked themes of the force of the Angara and the centrality of the dam to Soviet economic development are frequently mentioned. The familiar representation of dam building as a confrontation between unconscious, though active, nature and intentional and productive humanity is also underscored. In this struggle, human beings once again rise above nature, in their ability both to devise sensible plans and to implement those plans in an expeditious fashion. Discussing the logistics of changing the Angara's flow, one engineer notes that "we are changing [the motion of the river], but while the river itself laid out a course for itself over many thousands of years, we have been allotted only ten months for this matter."[34] This helps explain the constant sense of pressure that, as in many Soviet construction novels and in actual Soviet existence, underlies the behavior of the novel's characters. Their speed is an indication not of recklessness or carelessness but of Soviet humanity's intellectual power, intentionality, and superiority to nature—which drifts mindlessly for eons, its agency stunted.

The high point of the novel occurs when the dam is "closed," that is, finally completed. Much discussion, study, tales of damming other rivers by older workers, and watching of documentary films devoted to previous dams precede this moment. The excitement builds, and the sense of battle between nature and human beings becomes even more intense. Any thoughts of the Angara's aesthetic charms yield to a recognition of its essential enmity: "Rozhnov [one of the project engineers] wasn't thinking about the [river's] beauty now. Before him was an opponent, threatening in its frenzied force and merciless to the least confusion and mistake."[35] After the closing of the dam, the river quickly falls in line, however, ready to "serve devotedly and industriously the people who have subdued it."[36] Its anthropomorphosis here implies that like the Dnieper, the Volga, and many other rivers, the Angara, in its new guise as an agent, has become a useful Soviet citizen and member of the proletariat. The desired goal has been accomplished, speedily and efficaciously. Intelligent service has replaced wild destructiveness.

The Irkutsk or Baikal Hydroelectric Dam was only the first of several dams eventually constructed on the Angara. Of the dams subsequently erected, the one that claimed the most attention, and indeed represented the pinnacle of Soviet dam construction, was the largest in the Angara cascade: the gigantic Bratsk hydroelectric complex, located near the Padun Rapids, 666 kilometers from Lake Baikal, where the river narrows to 760–1,100 meters wide and pushes through cliffs 75–80 meters high. The major Bratsk Dam is 925 meters along its crest and reaches a height of 125 meters.[37] Leonid Brezhnev, among others, later enunciated Bratsk's technological and ideological significance: "Today, where there was once dense taiga, the Bratsk hydroelectric station, the greatest in the world, has arisen and the new city of Bratsk has grown up. It has become a symbol of the revolutionary maturity and courage of Soviet youth of the sixties."[38] If, as according to folkloric tradition, Baikal was the father of Angara, Bratsk, Soviet wits declared, was Baikal's grandson.[39] Bratsk was truly a match for Dneprostroi and implicitly for the TVA and similar capitalist-inspired projects. It was widely admired by Soviets and non-Soviets alike as a high modernist accomplishment and in literature as an opportunity for individuals to reveal themselves at their best.

In adherence to Soviet practice, many journalists, poets, and novelists visited the Bratsk construction site and sought to represent its significance in literary form. Years before the completion of the dam, the influential poet Aleksandr Tvardovskii (1910–71) produced *"Razgovor s Padunom"* (A Conversation with Padun, 1958), a poetic conversation between the poet and an anthropomorphized Padun Rapids. Here and elsewhere, such anthropomorphosis facilitates the representation of natural features as hostile opponents. Tvardovskii's Padun is a grouchy old man confident of his power and dismissive of human beings. Tvardovskii, however, cites both the human ability to triumph over great odds and human intelligence. He reminds Padun of the engineering accomplishment that has taken place at the Irkutsk Dam. In Tvardovskii's eyes, it is just a question of time before Padun will be forced into silence beneath the waters raised by a dam, in other words, rightfully and righteously drowned in service of the Soviet people.

46   THE HIGH MODERNIST HEYDAY OF MEGA-DAM CONSTRUCTION

Prose works also insist on the unquestionable virtues of human control of nature at Bratsk. Anatolii Pristavkin (1931–2008) is best known in the West for his glasnost-era wrenching portrayal of the internal deportation of the Chechens from the Caucasus to Siberia during World War II, *Nochevala tuchka zolotaia* (A Golden Cloud Spent the Night, 1987), years before wrote several stories and a long memoir, *Angara-reka* (The Angara River, 1958–77) that drew on his experience as a young man at Bratsk. Like the writers mentioned above, in *The Angara River* Pristavkin repeatedly emphasizes the struggle between inchoate nature and rational human beings. For Pristavkin, the very sight of the Angara evokes energy and force and the river's destiny as "a river of electricity."[40] The river will submit to human plans: "She will go, my dear, darling Angara, a beautiful, wild northern river, along a path that for now only exists on paper . . . houses, cities, settlements, a dam, industrial complexes, new factories . . . There will be everything, everything!"[41] In *"Zapiski moego sovremennika"* (Notes of My Contemporary, 1967), Pristavkin makes the Angara's current enemy status and socialist future even more explicit: "The wild waters of the Angara will hit against the wall of the 120-meter dam, they will die down for a moment, as if in amazement, and blinded, will rush inside in a frenzy in order to smash all this, to warp and carry it away. But the waters will be smashed to bits against the blades of the turbines, in order that they may slowly comprehend their laboring purpose and get used to working."[42] Once again, an anthropomorphized river will, it seems, acquire a proletarian persona and happily perform the duties of a Soviet citizen.

The most famous literary work produced about Bratsk was the narrative poem *Bratskaia GES* (Bratsk Hydroelectric Station, 1965) by Evgenii Evtushenko (1933–2017), for whom the poem may have represented in part a politically motivated atonement for expressing too-radical criticisms of the Stalin era a few years earlier at the height of the thaw that followed Stalin's death in 1953. *Bratsk Hydroelectric Station* includes a dialogue between an anthropomorphized Egyptian pyramid and the Bratsk Dam. The use of the pyramid calls to mind the contrast drawn between the Great Pyramid of Cheops and the Irkutsk Dam in Taurin's *Angara*, with its proud implication that a Soviet hydroelectric dam is superior to an Egyptian pyramid because of the socialist labor involved and because a dam is useful while a pyramid is not, a conclusion by no means unique to Evtushenko or even to Soviet

The High Modernist Heyday of Mega-Dam Construction   47

writers. Evtushenko's pyramid is congenitally pessimistic. When, in its mysterious eternal wanderings, it comes upon the Bratsk site teeming with people, the pyramid is certain that it is witnessing once again an instance of slave labor on a massive scale. In fact, human beings, according to the pyramid, are by nature slaves. The dam disagrees with the pyramid and embarks on a lengthy overview of Russian history that culminates with Lenin's accomplishments. Silenced by the dam's exhaustive account and convinced of the accuracy of its repeated claim that communists will never be enslaved, the pyramid vanishes.

The pyramid's disappearance prefaces a number of vignettes largely devoted to a range of workers at the Bratsk site and their specific backgrounds and attitudes. Reminiscent of Bezymenskii's old man, a local old woman indicates a willingness to see her home submerged so long as it is for the good of the people. An unmarried and pregnant young woman is restrained from drowning herself by "her" Angara: *"I krichala moia Angara: 'Kak ty moshesh' takoe, Niushka!'"* (My Angara cried out: "How can you do such a thing, Niushka").[43] Presumably, Niushka's personal difficulties fade in the face of her socialist commitment. An old Bolshevik who had been imprisoned as an enemy of the people during the Stalin era and forced to work on the postwar construction of a dam on the Volga now has his party card back and labors with enthusiasm at Bratsk.

In the final section of *Bratsk Hydroelectric Station*, Evtushenko discusses the tour he took of the dam's interior, a lovely excursion on the Angara, and, surprisingly, the beauties of nature. He reflects on the importance of art and political commitment. His thoughts conclude with a mystical apprehension of Bratsk's great importance: *"I bylo slyshno mne, kak GES gremit / v osmyslennom velichii-nad lozhnym / bessmyslennom velich'em pyramid . . . / Mne v Bratskoi GES mertsaiushchii / raskrylsia, Rossiia, materinskii obraz tvoi"* (And I heard the hydroelectric plant resound in its sensible greatness over the false, senseless greatness of the pyramids . . . In the Bratsk Hydroelectric Station, Russia, your shimmering maternal image was revealed to me).[44] This is indeed a lofty poetic assessment of a construction project. Reason has assumed physical form in the dam, and the dam represents Soviet Russia, the mother of the Soviet people, much as the Brooklyn Bridge symbolized modernity to Maiakovskii. The pyramids are but a symbol of a useless, politically mistaken past.

48   THE HIGH MODERNIST HEYDAY OF MEGA-DAM CONSTRUCTION

From Hoover Dam to the Tennessee Valley Authority

As in tsarist Russia and the Soviet Union, there was intense interest in the United States in the late nineteenth and early twentieth centuries in reaping the many benefits associated with big-dam construction—navigation improvement, flood control, irrigation, and hydroelectric power. The political challenges that confronted such development in the United States were complex. River basins typically encompassed multiple states, whose varying interests had to be delicately balanced. The issue of public versus private utility companies was also a matter of great controversy; concerns about private monopolies contended with accusations that excessive federal involvement in projects would lead to an incipient and dangerous socialism, a particular fear at the height of the Cold War but one that began to gain traction soon after the Russian Revolution of 1917. Within the federal government, two large and competing bodies were involved in riverine development: the United States Army Corps of Engineers, with its emphasis on navigation and flood control, and the United States Bureau of Reclamation (formerly the Reclamation Service), with its own emphasis on irrigation and, later, hydroelectric power. Initiating and implementing a large-scale project could require years of effort and lengthy arguments in the United States Congress.

Dams had been built before in the United States, but in the American imagination Hoover Dam, constructed between 1931 and 1936, at the height of the Great Depression, occupied a unique position analogous to that of Dneprostroi in Soviet visions (figure 2.3). The construction of Hoover Dam provided massive employment and helped offset Depression-era poverty. It also epitomized the heights, literally and figuratively, to which capitalist construction could rise. Situated on the Colorado River, on the border between Nevada and Arizona, Hoover Dam was so called, in honor of President Herbert Hoover, from 1930 to 1933 and was known as Boulder Dam from 1933 on.[45] It was renamed Hoover Dam by the United States Congress in 1947. The name Boulder Dam was ironically inaccurate. Since the beginning of the twentieth century, engineering investigations on the Colorado, increasingly under the auspices of the Reclamation Service, had considered first Boulder Canyon and then the slightly more southerly Black Canyon as possible sites for a dam that would help control floods, assist with irrigation, and provide hydroelectric power—especially

FIGURE 2.3. P. Hughes, *View of Hoover Dam and Part of Lake Mead, Nevada-Arizona Border*, photograph, May 17, 2013. Wikimedia Commons, https://commons.wikimedia.org/wiki/File:Boulder_Dam_-_A.jpg.

to Southern California. By the time Black Canyon was deemed the geologically superior site, the Boulder Canyon name had become entrenched in legislative documents.

Political resistance throughout the 1920s to the project's daunting size and cost eased after 1927, when horrific flooding on the Mississippi River convinced many members of Congress that large-scale river basin planning had decided merits. The Boulder Canyon Project Act received federal approval at the end of 1928. In 1931, under the auspices of the Bureau of Reclamation, the federal government, which committed to supplying materials, began to accept bids from contractors for construction of the dam. The winning bid was the result of a joint venture by six large western companies. The engineer in charge was Frank Crowe, who had established a significant reputation as a dam builder.[46]

Despite its associations with the New Deal, the Boulder Canyon Project was not originally intended as a public works project, but the desperate economic conditions of the Great Depression meant that huge numbers of unemployed workers immediately began to descend on southern Nevada in hopes of finding employment at the dam site. This led to a certain level of employment but also to very difficult living conditions, a situation the area's intense desert heat only complicated. Tellingly, an impromptu community that came into existence was known as Ragtown. Throughout the construction years, there was a struggle to provide decent conditions for workers. The Industrial Workers of the World (IWW), an international labor union with socialist links founded in Chicago in 1905, whose members were nicknamed the "Wobblies" and whose influence was in decline by the 1920s—partially because the American Federation of Labor considered its politics too radical—sought unsuccessfully to organize workers at the dam site. Working and living conditions, as well as pay, improved only a little during the construction period. There were also serious issues with discrimination, or what one writer has called "government-supported racism," against Asian Americans, African Americans, and Native Americans.[47] Discrimination involved access to both employment and decent working and living conditions. All of this was overlooked in the excitement of capitalist creation.

The Hoover Dam is more than 213 meters in height and 366 meters in length. Its width goes from more than 183 meters at its base to 14 meters at its crest. When the hydroelectric power plant at the dam came into full operation in 1939, it was the largest in the world. So much for Dneprostroi. Lake Mead, the reservoir behind the dam, still has the largest water capacity of any reservoir in the United States. In recent years, however, the reservoir's level has dropped significantly, in part because of ongoing drought conditions.

As occurred with Dneprostroi, a stream of journalists and writers visited the Boulder Dam site during and after its construction, and many of them made glowing statements about its significance. In a book on the Colorado originally published in 1946, Frank Waters (1902–95), who later had second thoughts about the Hoover Dam and was best known for his novel *The Man Who Killed the Deer* (1942), waxed enthusiastic: "Boulder Dam is the Great Pyramid of the American Desert, the Ninth Symphony of our day, and the key to the future of the whole Colorado River basin . . . it stands in its desert

The High Modernist Heyday of Mega-Dam Construction   51

gorge like a fabulous, unearthly dream. A visual symphony written in steel and concrete . . . it is the greatest single work yet undertaken to control a natural resource dominating an area of nearly a quarter million square miles."[48] Waters's enthusiasm was shared by, among others, the prominent journalist and author John Gunther (1901–70), who in his massive tome of regional essays, *Inside U.S.A.* (1947), called the Hoover Dam "one of the supreme works of modern man in America."[49] Not everyone was impressed. One of the few writers who did express criticism of the project relatively early was Edmund Wilson (1895–1972), who denounced the dangerous and unhealthy conditions in which the workers labored and rested. Wilson attributed the tolerance from above of such conditions to the rapaciousness of capitalism: "There was nothing to restrain the companies from resorting automatically and immediately to that systematic skimping, petty swindling and frank indifference to the fate of their employees which are necessary to provide officers with salaries and stockholders with profits."[50] Unlike most of his peers, Wilson had apparently paid attention to socialist propaganda, but he was a rare exception.

Although the construction of Hoover Dam did not result in the same volume of laudatory literary works devoted to Dneprostroi, two major novels were produced that told the story of the dam's construction in adulatory terms: the posthumously published *Boulder Dam* (1963) by Zane Grey (1872–1939) and *Big Red* (1966) by John Haase (1923–2006). The two conservative authors shared an admiration for the Hoover Dam and its engineers and a keen distaste for communism, unions, and anything that smacked of socialism. Haase's family had been refugees from Nazi Germany.

Best known for his novels set in the American West, most notably *Riders of the Purple Sage* (1912), Grey commanded extraordinary popularity with unsophisticated readers but did not always win the approval of the American literary establishment, in part because of his often pedestrian writing style. Grey was extremely prolific, though. At the time of his death, so many unpublished book manuscripts remained among his papers that publishers were able to release titles for decades afterward, hence *Boulder Dam*'s posthumous publication date of 1963.

Boulder Dam is set mainly in 1932 and, like many of Grey's novels, is a coming-of-age story. Lynn Weston, a young Californian from a wealthy family with some educational background in engineering, starts to work at the dam site in part because his well-to-do girlfriend has rejected him. Essentially goodhearted but initially lacking maturity, Lynn works in a variety of jobs at the dam, from steam shoveler to supervisor, and gradually hones both his moral and political sense—much like a good Soviet worker but in a different key. Along the way he falls in love with a sensible young woman, whom he rescues more than once from a white slavery ring. Lynn also rescues a fellow worker from drowning at the dam site and learns of and thwarts a communist plot to wreck the dam.

In Grey's representation, nature, and the Colorado River in particular, assumes an awesome and threatening guise reminiscent of the way nature is portrayed in the Soviet literary works discussed above. Grey sets the stage in the prologue to the novel, in which he provides a geological narrative of the birth and growth of the Colorado, which in *Boulder Dam*'s treatment takes on a familiar anthropomorphic and hostile character. It is "a rapacious and terrible river," "a sinister and irresistible stream," "mighty and insensate," with "teeth of sand."[51] After millions of years, the Colorado becomes the object of human intentionality and a wondrous and rational idea, "an idea born of the progress of the world, as heroic and colossal as the inventive genius of engineers could conjure, as staggering and vain as the hopes of the builders of the pyramids, an idea that mounted irresistibly despite the mockery of an unconquerable nature—and it was to dam this ravaging river, to block and conserve its floods, to harness its incalculable power, to make it a tool of man."[52] Grey checks several high modernist boxes in this brief accolade—the unquestioning belief in the value of progress and the need to make nature submit to human demands, the sense that nature in its wild agency is violent and unreasonable, and the adulation of engineers as agents of technology. Grey's engineers, however, are implicitly driven by capitalist, not socialist, visions.

Where Grey briefly departs from the high modernist script is when he mentions the pyramids. Instead of asserting the superiority of hydroelectric dams to the pyramids, as many authors do, he draws an implicit parallel between the aspirations of the pyramid builders and those of today's engineers. The reference to "the mockery of an unconquerable nature" introduces into the narrative an element of doubt about the ability of humans to tame nature completely. The prevailing tone is optimistic, but there is a vague sense that any victory over nature will not be permanent. Such doubts,

The High Modernist Heyday of Mega-Dam Construction   53

which Lynn Weston also periodically expresses in the novel, were alien to high modernism, with its rapt focus on the immediate future.[53]

Lynn responds immediately and wholeheartedly to the progressive task embodied by the dam project: "The tremendousness of that engineering project and the magnificence of its setting in the Black Canyon of the Colorado had struck Lynn with staggering force at his very first sight and conception of them. They had changed the direction of his life; they had set him at a man's job; they had been responsible for the gradual development of his character."[54] The young man does, however, harbor concerns about the ability of human beings to make the Colorado submit permanently: "In Lynn's secret opinion only the elemental forces that had given birth to this strange river could ever change its course or dam it permanently."[55] Perhaps nature does indeed possess great agency? Like Grey's narrator, Lynn too ponders the example of the pyramids: "The pyramids in Egypt, the Sphinx, the Appian Way—these constructions of man had lasted thousands of years beyond the lives of those who sought service out of them. But what was a few thousand years to nature?"[56] Such thoughts gradually diminish in the face of the grandeur of the construction site and the recognition of human ingenuity, particularly as embodied in the efforts of Carewe, the fictional engineer in charge of the project: "Was not the mind of modern man, heir of all the ages, the most marvelous of all developments? Was not Carewe, the chief engineer on this seven-year job, a mathematical and constructive genius who made the firmament his playground?"[57]

Carewe is a godlike figure to Lynn, who concludes: "The great engineer Carewe was supreme. He would complete Boulder Dam. Only nature and a hundred thousand years could destroy that work."[58] By the end of the novel, when the dam is complete, Lynn realizes that despite nature's ineluctable persistence, the human drive to create will also persist. He imagines Boulder Dam 500,000 years in the future: "Life had failed on the earth. Inscrutable nature had gone on with its work, patient, terrible and endless . . . But the earth, with its long past age of creation, its dreamers and builders who had passed on, was only a tiny globe in the universe. Other planets were evolving. And that divine Thing felt by Lynn in his vision had no beginning and no end. The spirit moved ever toward perfection and immortality."[59] Here then is a resolution of sorts of the dilemma of the eternity and agency of nature and the ultimate transience of human efforts. For Lynn, no single human victory

54   THE HIGH MODERNIST HEYDAY OF MEGA-DAM CONSTRUCTION

over nature will be completely lasting, but the human striving to wrest temporary victories from nature will be permanent. With its hint of religiosity, the tinge of mystical thinking exhibited here diverges from the unrelenting positivism of Grey's Soviet contemporaries, but Lynn's ultimate realization, that the human striving to dominate nature will never truly yield, provides an optimistic conclusion to *Boulder Dam*. Capitalism will be the winner, at least for the next 500,000 years.

As in the Soviet works discussed above, Grey casts the struggle to build a huge dam in explicitly military terms. Early in the novel, as Lynn and others approach the site and hear and feel dynamite blasts, thoughts of a battleground come immediately to mind: "'Boys, we're near the front now,' one individual said with a faraway look in his faded eyes. 'You said it, buddy. War!' ejaculated another."[60] Clearly, this is a war in which the participants are eager to engage.

Another struggle is being waged at Grey's Boulder Dam site, however, that constitutes the inverse of Soviet literary attempts to show the growth of a new labor ethic under the auspices of socialism. The living and working conditions Lynn encounters exemplify all the wonders of a benevolent capitalism. Lynn and his co-workers are provided with the most suitable clothing for their labors, excellent food, and clean, safe living conditions. Boulder City, the planned community for the workers, which on Lynn's arrival is still under construction, promises to become a "shining and model city."[61] There is no mention of Ragtown.

It is difficult to imagine why anyone would wish to tamper with this labor paradise, but Lynn soon discovers the existence of a communist plot to destroy the dam of which, ironically, Ben Brown—the worker Lynn rescued from drowning—is a member. Lynn plays a major role in thwarting the plot and heroically prevents an entire group of vicious cowards from damaging the dam. Lynn himself cannot understand the motives of Brown and his fellow conspirators: "Ruin the dam—make useless all the labor, the materials, the millions—and the sweat and blood—the lives of men who have died for it? . . . Oh, improbable, unthinkable! What a ghastly thing. There couldn't be men such dastards."[62] The explanation for this conundrum, eventually provided, is that men like Brown are ignorant dupes who fall prey to wily communist agitators (dastards) who parrot slogans about the emancipation of the working man and whose essential Americanism and devotion to

The High Modernist Heyday of Mega-Dam Construction   55

capitalist endeavors are suspect. The chief architect of the communist plot is the devious and obviously evil Sproul. At no point in *Boulder Dam* is there any sense that a legitimate basis for communist-inspired labor agitation might exist. Grey's Reds are cowards, cornered rats. In their mindless urge to destroy, they are the mirror image of the one-dimensional retrograde plotters who repeatedly surface in Soviet construction novels. Their assault on the American dream is equally unfathomable.

Haase's *Big Red* is reminiscent of *Boulder Dam* in both its plot and its politics. The historical Frank Crowe is an actual major character in the novel but in a heavily fictionalized version; details about his marriage, for example, do not correspond to historical reality. Haase's Crowe, like Grey's Carewe, is almost superhuman—an extraordinary engineer, penetrating strategist, master of the swift yet impeccably accurate decision, and tough yet caring human being. His driver is Hans Schroeder, a young man from a much less affluent background than Lynn Weston but similarly decent and with an equally touching love interest. Hans's brother, Herman, is a different story. Bitter and dissatisfied with life, in part because of the loss of a leg in a childhood accident and its replacement by an awkward wooden one, he has fallen prey to the manipulations of the IWW. The local leader of the IWW in the novel is Devere, a World War I veteran who is also embittered, in part because of injuries sustained during the war; his testicles were shot off, his masculinity compromised. As in *Boulder Dam*, much of the action in *Big Red* revolves around an ultimately foiled IWW plot to sabotage the dam.

Crowe explains the title *Big Red* early in the novel: "The locals call that river [the Colorado] Big Red, and it's the toughest, meanest river in the world."[63] The impression the Colorado makes on Crowe's fictional wife, Elle, is even more ominous: "Elle was hypnotized by that constant brown body glistening in the summer heat, and for the first time sensed its strength and its relentlessness and its utter silence, which made it as frightening as a jungle cat, immobile and poised for the kill."[64] For Crowe and others, the river is indeed an enemy adversary with violence as its object. The seemingly obligatory war metaphor is enunciated early in the novel when the stark landscape at Boulder Canyon reminds another engineer of "a deserted battlefield, one in which the battle was really about to begin."[65] For Crowe, the river is his "biggest adversary. The Colorado river . . . had not yet been tamed and might yet prove the victor."[66] In the end, though, the adversary is definitively conquered by completion of the dam, and Crowe triumphantly declares: "Big Red, crafty, forceful, never-yielding adversary until we yoked it barely a few months ago."[67] As in many literary works and in contrast to Grey's *Boulder Dam*, the victory over the river is absolutely unquestioned in *Big Red*. The Colorado's yoking recalls the chaining of Bash's Dnieper.

The politics in *Big Red* are presented in black-and-white terms reminiscent of Grey's *Boulder Dam* but more brutally and crudely. Haase's Crowe is sensitive to his workers' physical needs and aware of the existence and problems of Ragtown, and he makes attempts to address them. He gives serious attention to issues of safety and cleanliness. His attitude toward the dam workers is ultimately intensely paternalistic, though. Crowe very much views himself as the one who is in charge, or as he puts it, the one who thinks "*I* get paid for thinking. But when *they* think . . . we're in trouble, because I *think* better, and that's *my* job."[68] This attitude helps explain Crowe's utter lack of patience for the IWW or any kind of strike activity on the part of the dam workers. "I'll have no labor shit on this dam," he says repeatedly.[69] Crowe has no tolerance for communists or, for that matter, for any kind of government interference in his labor practices: "There were times when Frank truly questioned his system of government, whose laws allowed these agitators to move about in freedom, when Crowe felt they should be strung up and let dangle so other men could go about their work unfettered. And now he was angry that the U.S. Senate had set up an investigative committee to 'study conditions at the dam.'"[70] Frank Crowe, the ultimate heroic engineer, knows best, in other words. This attitude helps explain his preference for Herbert Hoover over Franklin Delano Roosevelt, whom he suspects of wanting to foster socialism. Some of his partners, however, are able to convince Crowe to at least consider the potential wisdom of Roosevelt's approach to the problems produced by the Great Depression.

Herman Schroeder and Devere fully confirm Crowe's suspicions about communists. In the case of both characters, there is a clear link between their physical handicaps and their embrace of communism. Both in a sense are attempting to compensate for feelings of inadequacy, Herman because he is constantly called a cripple and Devere because the loss of his testicles symbolizes to him a loss of manhood. Of Herman, Haase writes: "He was an easy mark for the IWW, which thrived on men like Herman, since their aim, the destruction of capitalism 'by force,' matched on a grander scale the same revolt they experienced in themselves."[71] Devere is equally psychologically damaged: "He thrived on strife and tension, his mind filled with slogans, theories, to the point where he subconsciously no longer remembered *what* cause he was fighting, or why, or at what price. He was a political eunuch, the best kind of radical, who moved when told to move, no matter what the cost or what the unreality."[72] The loss of his testicles, which Devere frequently mentions, has led him to hatred and thence to communism; perceived emasculation makes him vulnerable to what Haase presents as unmanly socialist ideas. Devere is intelligent but not as intelligent as Crowe. He is also cruel and lacking in empathy. His denunciation of Herman in the middle of an argument as a "crippled little Hun" leads Herman to pistol-whip Devere and then shoot himself in the mouth.[73] So much for socialist camaraderie.

The heroic vision of the construction of the Hoover Dam by Grey and Haase was later revised by some American authors. In Phyllis Barber's novel *And the Desert Shall Blossom* (1991), which draws to a great extent on historical details about the project, the corruption, exploitation, and poor and dangerous working conditions at the dam receive extensive attention. Even some writers who had earlier expressed enthusiasm later developed doubts. As mentioned in chapter 1, in 1984 Frank Waters acknowledged in a new preface to *The Colorado* that perhaps he had overestimated the human capacity to conquer nature and the wisdom involved in the efforts to do so exhibited by the dam. These kinds of doubts began to emerge more frequently in the late twentieth century.

Other dams were later built on the Colorado River, but immediately after Hoover Dam was completed, engineering and public attention shifted to the Columbia River Basin—which includes part of British Columbia and most of Washington, Oregon, and Idaho, as well as portions of several other states.[74] The struggles involved in building dams and supposedly taming rivers generated more literary interest than did the calmer aftermath. The Columbia evoked great interest. An extraordinarily powerful river, dumping more water into the Pacific Ocean than any other river in the Western Hemisphere, its annual flow is almost ten times that of the Colorado. Today, nearly half of the hydroelectric power generated in the United States is produced by dams on the Columbia and its tributaries.

Throughout the 1920s, competing political forces struggled to resolve the question of development of the Columbia River. Private power companies

FIGURE 2.4. *Grand Coulee Dam on the Columbia River, Washington*, photograph, 1941. Library of Congress, Washington, DC, http://loc.gov/pictures/resource/fsa .8e01538/.

resisted perceived government intrusion, once again playing on fears of socialism—fears underscored by the existence of the Soviet Union. The states of Oregon and Washington pushed their respective interests. The United States Bureau of Reclamation and the Army Corps of Engineers fought for control over decisions. In July 1933 construction of a low dam at Grand Coulee under the auspices of the Bureau of Reclamation was approved. The plans were subsequently modified to permit construction of a much higher dam that would take full advantage of the site's awe-inspiring geological features (figure 2.4). Coulee refers to a massive gulch formed during the Ice Age. The site, bordered by steep cliffs, afforded an ideal site for a dam. When completed, Grand Coulee was 168 meters high and 1,592 meters long; its width at the base was 150 meters and at the crest, 10 meters. The construction of Bonneville Dam—another important dam on the Columbia—under the auspices of the Army Corps of Engineers received federal approval in September of the same year. The Public Works Administration, created as

FIGURE 2.5. Al Aumuller, *Woody Guthrie, Half-Length Portrait, Facing Slightly Left, Holding Guitar*, photograph, 1943. Library of Congress, Washington, DC, https://www.loc.gov/resource/ppmsca.74704/.

part of the 1933 New Deal, helped fund construction of both dams. In 1937 the Bonneville Power Administration (BPA) was established, as a division of the United States Department of the Interior, to act as an agent for the sale and distribution of hydroelectric power, initially from the Bonneville Dam and later from Grand Coulee and all other dams in the Pacific Northwest.

Bureaucrats at the BPA were anxious to propagandize the benefits of the construction of large dams on the Columbia River and soon produced a documentary film, *The River*. Stephen Kahn, a canny BPA official, decided that a subsequent film, devoted above all to the Grand Coulee Dam, would benefit from the inclusion of more engagement by a ballad maker. The ethnomusicologist Alan Lomax, director of the Archive of American Folk Song at the Library of Congress in Washington, DC, recommended the popular and immensely influential singer and songwriter Woody Guthrie (1912–67) (figure 2.5).[75] Guthrie received a month-long contract from the BPA in the late spring of 1941 and wrote twenty-six songs about the great happenings on the Columbia River in the space of a few weeks. Three were included in the film *The Columbia*, which was released in 1949. Other songs appeared in albums and songbooks.

Guthrie was captivated by the Columbia River Basin Project. As Alan Lomax put it: "He saw the majestic Grand Coulee Dam as the creation of the common man to harness the river for the common good-work for the jobless, power to ease household tasks, power to strengthen Uncle Sam in his fight against world fascism [note the spirit of competition here]. Listening to the whine of generators in new factories, looking up at the shining grid of towers and power lines, Woody opined that maybe 'the whole damn country ought to be run by electricity.'"[76] Ironically, Lomax goes on to compare Guthrie to Maksim Gorky in his appreciation of the working class. Unlike the conservative Grey, for example, Guthrie did have decidedly leftist leanings.

Kahn arranged for a chauffeur to drive Guthrie around the Columbia River Valley so the singer could "capture 'the thoughts and ideals' behind the Bonneville and Grand Coulee Dams."[77] He also warned Guthrie to monitor his politics—that is, not to introduce offensive leftist elements into his lyrics. On his tour up and down the valley, Guthrie saw squatters' camps, apple orchards, wheat fields, and the Bonneville and Grand Coulee Dam sites. He responded with unfeigned excitement. Kahn declared: "'Guthrie was thrilled with the prospect. He saw it was more than a power or a reclamation or a navigation project, but something that could touch the lives of the people of four or five states and set a pattern of how democracy could function in this country with the government doing something constructive to improve the conditions of the people.'"[78] In other words, like his Soviet literary counterparts, Guthrie appreciated the socially transformative implications of dam construction, but in this case they were to be cast as democratic, not socialist. He succeeded admirably in his assigned task. The oral historian and radio broadcaster Studs Terkel (1912–2008) claimed, "A Washington State senator said that any one of these Woody songs was worth a dozen legislative speeches in getting things done."[79] In 1966 Guthrie received a citation from the Department of the Interior for the songs he had composed for the BPA. Not everyone was impressed. The journalist and author Blaine Harden (b. 1952) later took a more cynical view of Guthrie's efforts, terming them "Bureau of Reclamation press releases dressed up in cornpone rhymes."[80]

Many of Guthrie's twenty-six songs about the Columbia River project were set to familiar melodies, like "Old Smokie" and "Brown's Ferry Blues." Lomax points out that this is a traditional ballad composition practice intended to quickly engage listeners' attention.[81] A possible measure of the

The High Modernist Heyday of Mega-Dam Construction   61

effectiveness of this technique is that one of the songs, "Roll on Columbia, Roll On," was eventually designated the official state folksong of Washington state. The song's melody drew on "Goodnight Irene," which had achieved great popularity in the version recorded by Huddie "Lead Belly" Ledbetter in 1933.

The lyrics to "Roll on Columbia, Roll On" and the similarly titled "Roll, Columbia, Roll" touch on a number of themes that are repeated throughout Guthrie's corpus of BPA-inspired songs. The navigational improvements made possible by the Bonneville Dam locks are mentioned in "Roll on Columbia, Roll On," as is the industrial development that will be enabled by the hydroelectric power generated by the Grand Coulee Dam. Guthrie calls the Grand Coulee "the mightiest thing ever built by a man."[82] The song's claim that the power of the Columbia River "is turning darkness to dawn" points to the benefits of electric light (Lenin would have approved). In "Roll, Columbia, Roll" the factories that will draw power from the Grand Coulee are mentioned again, and electricity's superiority to coal and oil is touted. In this song, the pre-dam Columbia is described as "wild and wasted"—a traditional modernist characterization of rivers—and is urged to submit to human direction: "While you're rambling river you can do some work for me." In other words, once again nature must submit to human beings.

"The Biggest Thing That Man Has Ever Done (the Great Historical Bum)" provides a historical perspective on the accomplishment represented by the Grand Coulee. The pyramids receive their obligatory mention and an entire series of North American achievements is listed—the victory over the Germans in World War I, the battles won during the Revolutionary War, the triumph of freedom represented by the Civil War, and architectural monuments like the Empire State Building and the Golden Gate Bridge. The Grand Coulee surpasses all of these and more: "Three times the size of Boulder or the highest pyramid . . . The Coulee is the biggest thing that man has ever done."[83] Similar thoughts are expressed in the "Ballad of the Great Grand Coulee" and "The Song of the Grand Coulee Dam," often using the same phraseology. "Guys on the Grand Coulee Dam" focuses on the employment opportunities offered by the construction of dams on the Columbia and lauds the varied efforts of the dam workers.

In their totality, Guthrie's lyrics bear out Lomax's comparison of the folksinger with Gorky. By the time he composed the Columbia River songs,

Guthrie had become "evangelical about electricity and industrial production, determined henceforth to associate folk music with modernity rather than with the antique ruralism with which it was so often associated."[84] Like Soviet communism, the New Deal promised to create a brave new world—one of agricultural plenty, industrial successes, and a better daily existence for all but in a democratic, not a socialist, guise. Guthrie was thrilled to give this new world his endorsement, and he did so in captivating songs.

Once World War II began, the dams on the Columbia River required little boosterism. The hydroelectricity they provided greatly facilitated ship construction and the production of aluminum needed for aircraft. Hydropower from the Columbia was also a key factor in the choice of the Hanford site in south-central Washington state for part of the Manhattan Project efforts. The first nuclear reactor built at the site was used to manufacture plutonium subsequently employed in the atomic bomb dropped on Nagasaki, Japan. President Harry Truman later declared, "Without Grand Coulee and Bonneville dams it would have been almost impossible to win this war."[85]

After the war ended, Gunther visited the Columbia River and devoted an entire chapter of Inside U.S.A. to his impressions. While Americans in the 1930s and 1940s were rarely inclined to mention the massive dam projects in the contemporary Soviet Union, Gunther was clearly pleased to note that although "in prewar times the Dnieper installations in Russia were generally considered the most powerful of their kind on earth," the Grand Coulee's capacity was almost three times that of Dneprostroi.[86] So much for Soviet efforts and socialist political thinking.

Gunther was also impressed by and describes at length the fish ladders at the Bonneville Dam. As he observes, there had been some concern before the Columbia River dams began to be constructed that they would have a deleterious effect on the anadromous salmon, preventing them from spawning and thus ruining the important fishing industry on the river. From Gunther's point of view, the ladders were an unqualified success ("among the most ingenious things I have ever seen").[87] At the Grand Coulee, where fish ladders were not an option because of the dam's tremendous height, Gunther was equally impressed by the artificial insemination practiced on salmon and the fact that juvenile fish spent their youth in hatcheries: "Moreover—believe it or not—these fingerlings are so conditioned that, when they grow up and return upriver to spawn, they are content to do this in the substitute hatcheries, and so do not attempt to cross the Coulee barrier."[88] Time would show that much of Gunther's unmitigated confidence in such artificial measures taken to enable the growth of fish was misplaced.

The New Deal–era project of the 1930s and 1940s most clearly associated in the United States with the notion of societal transformation, not just water control, was the Tennessee Valley Authority.[89] The TVA is federally owned and affects seven southern states. It was authorized by the US Congress in 1933 to fulfill multiple purposes—navigational improvements, flood control, generation of hydroelectricity—and, more broadly, to foster agricultural and economic development of the Tennessee Valley, one of the poorest and most environmentally challenged areas of the United States, largely because of massive erosion. The TVA was a modernization project par excellence, intended not only to provide electricity, irrigation, and employment but to educate farmers on ways to improve and increase agricultural production. In 1934, President Franklin Delano Roosevelt enunciated his vision for the TVA: "There is a much bigger situation behind the Tennessee Valley Authority. If you will read the message on which the legislation was based you will realize that we are conducting a social experiment that is the first of its kind in the world, as far as I know, covering a convenient geographical area—in other words, the watershed of a great river. The work proceeds along two lines, both of which are intimately connected—the physical land and water and soil end of it, and the human side of it."[90] Such thinking is in many ways reminiscent of Soviets' hopes for their multiple cascades of dams. The TVA was ultimately responsible for the construction of twenty-nine hydroelectric dams, as well as several coal-fired power plants and nuclear plants. Today, it is the country's largest utility. More than 125,000 valley residents needed to be moved to accomplish this feat, but implicitly that move was more than justified.[91]

The TVA was initially heralded as an unqualified success. This was due in part to the astute propagandistic efforts of David E. Lilienthal (1899–1981), chair of the TVA from 1941 to 1946 and an untiring booster of the project. As mentioned in chapter 1, Lilienthal's treatise *TVA: Democracy on the March* (1944) and his later consulting efforts worldwide played a huge role in establishing the global recognition and admiration the TVA garnered while trying to avoid any suggestion of parallels with similar Soviet projects. In recent decades, however, the actuality of the success and efficiency of the

TVA have been called into question by a range of analysts. Yet the TVA continues to serve as an icon of development. In the early 2000s, for example, Abdelrahim Hamdi, the former Sudanese minister of finance, drew comparisons between the TVA and the massive dam construction taking place in Northern Sudan.[92]

Despite the worldwide excitement caused by the TVA, especially in the 1940s and 1950s, the project did not evoke a wealth of literary response. In the United States, the heyday of enthusiasm for dam building had already begun to fade. Two novels that did emerge were Eleanor Buckles's *Valley of Power* (1945) and Borden Deal's *Dunbar's Cove* (1958). The latter was one of the works adapted by the screenwriter Paul Osborn for the film *Wild River* (1960), directed by Elia Kazan. Deal (1922–85) was a prolific novelist, most of whose novels are set in the 1950s American South. Buckles worked primarily in secretarial and office capacities; *Valley of Power* was one of her very few publications. An assistant for a time to Thomas Wolfe, Buckles later worked for the TVA. *Valley of Power* reflects detailed knowledge and approval of the many agricultural projects associated with the TVA.

Valley of Power focuses on the final family removal efforts necessitated by the reservoir filling that followed the completion of the fictional Spaulding Dam, which is located in the mountains near Knoxville, Tennessee. The hero of the novel is Root Jonas, a young engineer who decides to assume the seemingly unrewarding post of family remover from lands to be flooded because he recognizes that the process of removal is central to the success of the TVA as a socially transformative movement. "I think I can bring the mountain people and the dam together," he hopes.[93] Root is as idealistic and driven as many young Soviet experts. In the course of the novel, he realizes that he needs to find a uniquely satisfying replacement of their old existence for each resistant family: "He had to play the hopes of the future against the strength of memories. He had to sell the mountain people a conception of what was to come, in exchange for their past. The pull of the past was strong. The promised future must be bright to compete with it."[94] With one exception, he is able to accomplish this idealistic yet paternalistic goal: the creation of a sense of agency and purpose on the part of those forced to move because of dam construction.

The dam under construction in *Valley of Power* is an unqualified high modernist beacon, a triumph of rational human beings over irrational nature.

The High Modernist Heyday of Mega-Dam Construction   65

Root's first, nighttime sight of the dam provides an inspiring vision: "As they rounded the turn, there was a blaze of arc-lights flaming toward them from the valley. It was beautiful, breathtaking, incredible. It was the dam . . . He sat there spellbound, looking down at that shimmering wall of light spanning the gap between the mountains . . . It was the miracle of construction all over again. It was a collection of human beings, maximum height about seventy-five inches, who were taking hold of a vast river system and systematically putting it under control."[95] Later in the novel, the dam stands firm against storm-produced floodwaters, a testament to human victory over nature.

Buckles lavishes much attention in *Valley of Power* on the various benefits attendant upon the TVA. The widespread erosion that very visibly marked the Tennessee Valley by the 1930s and contributed to a decline in agricultural production and an increase in wretched poverty is frequently mentioned. One of the TVA initiatives intended to counter these negative developments was the establishment of model farms, where local farmers could see for themselves how much could be accomplished through the use of electric-powered machinery and the judicious employment of the latest irrigation and fertilizer methods. In Buckles's novel, examples of this initiative are described at length. No comparison with Soviet efforts is suggested, but such comparisons cannot help but come to mind.

Buckles does not shy away from treatment of one of the more contentious aspects of the TVA and of many other dam construction projects worldwide: the relocation of coffins that will otherwise be flooded. Such displacements understandably often evoke painful, even violent, emotions in family members. Buckles describes such eruptions with sympathy but implicitly gives the final word on the subject to Abel Ballard—a young man who is drawn to the vision of progress represented by the TVA, eventually takes a construction job at the dam, and acquires an increasingly mature and steady persona. Early in the novel, in a very public altercation with his father, Purse Ballard, Abel defends the grave removal by the TVA personnel: "They know how to do things, and if anybody ought to be ashamed, it's the ones who stand in their way!"[96] Abel's perception that the TVA is a force for the better is shared by his wise Aunt Loos, who tells her niece Sandrey: "You've got to figure if maybe what's a nuisance to a few of us might be good for a lot of other people that we've never had a look at. And in the long run, maybe it's good for us, too."[97] As has been seen before, the notion of the need for self-sacrifice by the few for the good of the many appears frequently in rhetoric surrounding dam building, both socialist and capitalist. Here it is presented as a good thing.

The one character who can never truly come to terms with the dam and the need to move is Purse Ballard, who is portrayed as almost pathologically stubborn and self-righteous. By the end of the novel Purse, however, is tormented by doubts about his position and finally goes to see the dam. The dam's immensity and seeming indestructibility stun Purse, but a possible change in his thinking is stifled when he is suddenly killed in an accident at the site of the dam. "He saw the dam was right, and to hang onto his land would be a wrong thing," declares his son,[98] but Purse's death saves him from admitting his mistake. It also saves him from witnessing the blossoming relationship between his daughter and Root Jonas, a relationship that implicitly symbolizes the reconciliation of the TVA and the mountain people. *Valley of Power* ultimately confirms what a farmer from South Dakota tells a local at one point in the novel: "I don't guess you folks think about it that way, but what you're doing down South here can be a pattern for all the river valleys in the world."[99] David Lilienthal would have agreed wholeheartedly.

The plot of *Dunbar's Cove* recalls that of *Valley of Power*. The widower Matthew Dunbar has five young adult children and an aging, decrepit father. In the course of the novel one son goes to work for the TVA, another goes off in search of his wayward wife, and the third dies in a freak accident. The elder of Matthew's two daughters falls in love with and ultimately marries Crawford Gates, a TVA official. Like Purse Ballard in *Valley of Power*, Matthew is increasingly obstinate in his refusal to consider selling and leaving his farm, despite its imminent destruction by the rising waters caused by the completion of the fictional Chickasaw Dam. He is only convinced to do so at the very end of the novel, when Crawford locates another, as yet undeveloped cove for Matthew's consideration. Root Jonas's tactics in *Valley of Power* work in *Dunbar's Cove* in Crawford's efforts with Matthew Dunbar.

In telling his story, Deal drew on the history of Alabama's Guntersville Dam on the Tennessee River, authorized in 1935 and opened in 1939, whose construction necessitated the relocation of more than a thousand families. Throughout *Dunbar's Cove*, Matthew and Crawford argue repeatedly and at length about the pros and cons of the TVA. Matthew's desire to remain where his family has labored for generations is presented with a certain sympathy, but there is no doubt that the Chickasaw Dam represents progress and a better life for the region's inhabitants. The novel's opening paragraph presents a traditionally high modernist view of the Tennessee River as an object of natural chaos to be controlled by thoughtful human efforts: "This is the river . . . drunk sometimes on flood water . . . It can be quiet, too, though not peaceful for the violence lies always beneath the quietness. It is not a blue river, not yet, but it will become so, for this river is to be tamed and civilized as no other river in the history of civilization has been tamed and civilized. This is the Tennessee."[100] In contrast to the river, the Chickasaw Dam is a highly aesthetic, almost spiritual creation: "The building of a dam is like music; there is the overture, the beginning theme, the tentative scratch and start at the earth; there is the orchestration of sound and effect, the building of melodic lines, the sudden clash of crisis . . . It is an enduring symphony, poured and formed and solidified to the sight of the eye."[101]

In his discussions with Matthew, Crawford promotes the TVA as a project that will improve local existence in a multitude of ways—by improving navigation, enabling the creation of co-ops to bring in electricity, putting an end to malaria, overcoming widespread erosion, and providing the opportunity for gainful and productive employment. "You talk about that TVA like it was God," Matthew accuses Crawford, and Crawford retorts: "TVA is going to do more for this country than God ever did."[102] This, too, is a familiar idea. In Crawford's eyes, the TVA offers a vision for collective success as opposed to individual hardship: "It's a new idea, I'll admit, that a farming man can live an easy, town, kind of life, with electricity to do his work for him, and indoor toilets, and tractors to take the toil out of his life. That his land can be held against the erosion of wind and rain and protected against the tearing away of floods. That his crops and his produce can go cheaper to market with river transportation all up and down the whole length of the river . . . Dunbar's Cove is an idea for one man, while TVA is for the whole country."[103] The TVA, declares Crawford later in the novel, "is a corporation with a soul . . . TVA's got a conscience and a mission."[104] It is a testament to the power of human agency, which is indeed in Crawford's eyes superior to divine thinking: "There's one thing that God forgot . . . That dam—men had to add it to the river, to make the river serve the purpose it was intended for. All up and down the river and its tributaries, the water is controlled in its flow, it's made to work at giving forth power for the benefit of the people."[105] The young man's faith in the TVA is borne out when massive flooding in the area is prevented in early 1937. Clearly, subduing nature is the right thing to do.

The attitudes expressed in *Valley of Power* and *Dunbar's Cove* were shared by John Gunther. In *Inside U.S.A.* he devotes significant space to the TVA. He expresses no love for the Tennessee River, calling it "an obstreperous ugly river with an angry history" and emphasizing its wastefulness. The TVA would reform the river: "The wasteful, dangerous giant of the Tennessee was going to be put to work."[106] Remember the Dnieper. What most impressed Gunther, however, was the vision of transformative regional development embodied in the TVA and the project's ability to stimulate faith and loyalty in those involved: "People are happy because they are doing something creative, something bigger than themselves."[107] Gunther is certain that the TVA will offer the world a productive model for river and valley development. His convictions matched those embraced by many American technocratic and political elites in the 1940s and 1950s, as well as analogous Soviet figures.

The excitement and enthusiasm stimulated by big dams worldwide has by no means vanished, but in recent decades it has been dramatically undermined and qualified. It is difficult to imagine the unreflective composition today of a work like *Energy, Bratsk Hydroelectric Station*, "Roll on Columbia, Roll On," or *Big Red*. The explicit and implicit unqualified confidence in progress reflected in such works has suffered assaults and strenuous interrogation from too many directions—environmental, political, economic, and social—in both capitalist and socialist thinking.

Despite their very different political systems and the growing importance of the Cold War, in the mid-twentieth century the Soviet Union and the United States shared many cultural attitudes. An unwavering belief in the possibility and benefits of technological advances and their ability to defeat wild rivers, accompanied by an equally strong conviction that technology could facilitate societal transformation, informed the thinking of many Soviet and North American politicians, bureaucrats, engineers, and ordinary citizens. In both countries too, there emerged a keen sense that their progressive agenda and its implementation could offer a model for development in other nations around the globe. Both nations pushed their models worldwide, driven in part by the passions of the Cold War.

The High Modernist Heyday of Mega-Dam Construction 69

Large multipurpose dams encapsulated the dreams of high modernism like few other structures. As described in the novels, stories, and poems discussed in this chapter, dams provide a highly visible symbol of the human ability and right to subdue wild nature, to literally force nature to conform to an essentially artificial but nonetheless desired and desirable human project. In the literary works considered here, the immensity of such projects is more than justified by their assumed ability to create a bright future in which people will prosper, both physically and psychologically. A sense of the promise of this bright future heightens a communal feeling of involvement in an endeavor much greater than oneself. Call it socialist, call it democratic, in a sense it does not matter—the joy is the same. A sense of rapturous optimism and awe at wondrous accomplishments permeates the vast majority of mid-twentieth-century Soviet and American literary treatments of massive dam construction. Reading these works today, one is drawn into a cultural milieu in which human progress appears both beneficial and inevitable, especially to the young. Development, it appears, has no ugly underside, and ongoing settler colonialism has no risks, especially where mega-dams are concerned. There were no worries about those displaced by dams or possible negative environmental consequences. Those concerns would be expressed later.

3

Displacement and Alienation of Peoples Worldwide

In the decades that marked the heyday of dam building worldwide—an era of enthusiasm and excitement that in some countries unthinkingly persists—little literary or other attention concerned the impact on those who had to leave their homes, often under duress, to make way for mega-dams and the huge reservoirs that backed them. As we have seen, any thought the displaced received often took the form of familiar and comfortable assumptions about the need and willingness of the few to sacrifice for the good or betterment of the many, as well as an idealistic belief that the dispossessed would share in a brilliant electrified and irrigated future. The question as to whether those forced to move, often poor or Indigenous populations, would truly benefit rarely arose in the early twentieth century. Gradually, though, a dawning realization emerged in some quarters that displacement—far from representing a glorious new beginning—could herald alienation, impoverishment, a lasting sense of cultural and social loss, and problems manifesting agency in a new world. Literature is able to personalize and bring to life these developments and has done so with a vengeance.

https://doi.org/10.5876/9781646425976.c003

Authors around the world, in Russia, the United States, Egypt, India, and China, have addressed these painful issues in a wide variety of novels and stories—works in which the problems seem to grow over time rather than diminish. From Valentin Rasputin's (1994) poignant evocations of the trauma of elderly Siberian villagers uprooted because of the construction of dams on the Angara River to Idris Ali's (1998, 2007) rendering of Nubian social and economic upheaval because of the building of the two Aswan dams to Li Miao Lovett's (2010) portrayal of the horrific impact of the Three Gorges Dam on the livelihoods of the poor, writers with diverse political beliefs have produced a body of literature that leads the reader to confront the growing human costs of assumed progress on vast numbers of poorly educated workers and peasants. This chapter seeks to delineate in precise terms what some of those terrible costs are, how they have been realized in several of the countries in which they have occurred, and the important role writers of fiction have played in bringing them to our attention. The wealth of such writings underscores the significance accorded the problem of relocation.

In chapter 1, I mentioned several consistencies associated with relocation of peoples because of dam construction. To reiterate, dams are rarely built in heavily populated or affluent areas. Rather, dam sites are generally located in rural regions where the population lacks political clout and has difficulty expressing agency. Often, such populations are members of Indigenous or ethnic minorities who have little in common socially or otherwise with a dam's planners and beneficiaries. They are thus easier to dismiss as abstract and unenlightened beneficiaries of the supposed technological progress the dam will bring.

In the early decades of big-dam construction, as mentioned, the experts included in the planning process were often limited to engineers and geologists, the propagandists to politicians. It was only later in the twentieth century that the wisdom and desirability of enlisting social scientists with a more knowledgeable appreciation of human issues of relocation became increasingly apparent, albeit often only grudgingly admitted. The reluctance to give much attention to such issues has several sources. For the actual planners and builders of big dams, the combination of "a messianic fervour" and implementation of what constituted "an experimental technology" does not readily tolerate expressions of caution or reservations about the true value of mega-dams from those outside their realm of direct expertise.[1] In addition, perhaps to state the obvious, the dam industry has a vested financial interest in building dams;[2] big dams mean big money. As a partial result, estimates of costs and clarification of the particulars of those costs have sometimes been sloppily done and facilitated a move to quickly implement large-scale projects. Analysts have noted that construction cost-benefit calculations are frequently "incorrect or unsubstantiated."[3] Similarly, the costs of relocation are poorly figured and drastically underestimated, deliberately or thoughtlessly. This is by no means novel information; it has been apparent for decades to those who choose to pay attention.[4]

The well-known appeal of large dam projects to politicians and governments encourages dam industry inclinations to produce overly optimistic projections of the costs and benefits of dams. As noted earlier, dams are eye-catching symbols of modernity, as well as creators of at least temporary employment; as such, they exert particular appeal to politicians everywhere. This contributes to a willingness to downplay the potential human costs big dams may entail. As Edward Goldsmith and Nicholas Hildyard observe: "Put bluntly, few governments are willing to increase the funds allocated for resettlement if doing so threatens the economic viability of a project. And in the majority of cases, that is precisely what would happen."[5] The result can be a vicious cycle of financial shoddiness and dismissal of human concerns. A conviction that big dams will mean big money for some can accompany ignorance or drastic lack of concern regarding the human costs of such endeavors.

Cost-benefit calculations associated with large dams do not make errors or evade close examination only regarding relocation of peoples. Errors often emerge about construction costs and the potential economic benefits of dams—the generation of hydroelectric power and agricultural advantages derived from irrigation, for example. With the significant funding surrounding it, corruption easily and rapidly makes its way into the dam-building process. Devastating environmental consequences that result from large dams are often overlooked, unrecognized, or ignored. Clearly, all these problems have in turn a human impact that may be difficult to separate from the direct consequences of relocation. Chapter 4 treats environmental damage associated with dams as represented in literary works in greater detail. This chapter will focus more directly on the effects of relocation on peoples and the aftermath of such upheaval as portrayed in literature around the world. As will be seen, such impacts have been growing at a rapid rate, and literature has done an excellent job of humanizing the devastation—physical and psychological—large dams often bring.

One of the most detailed overviews of the various elements of resettlement was an early one provided in the 1960s by Neville Rubin and William M. Warren (1968) in a study of dams in Africa. At the beginning of their discussion, the authors point out the essential complexity of the planning process itself: the physical design of the resettlement area, including not only housing but associated infrastructure; the retention and development of social institutions and their staff; and the complex details of the actual move.[6] Before the move, insufficient research may be done on hydrological and social factors. Delays in the move last for years, causing in the interim economic and other hardships, while planning of the actual move—once it actually takes place—may occur in a hurried and careless manner. The psychological stress the relocated suffer may receive insufficient attention. Inadequate involvement of the relocated in the planning process may exacerbate such stress; the tendency of many bureaucracies to engage in top-down tactics can be very detrimental. Costs of resettlement and compensation for lost property or use of property are frequently ignored or underestimated. While official property owners may receive monetary awards, albeit inadequate, tenant farmers and extended family members frequently receive no compensation. The economic impact of the loss of common lands and the use of rivers themselves, for fishing or other purposes, is often dismissed; this also affects downstream river dwellers, who may not have to move but whose livelihoods are affected by the creation of dams. Once dams are constructed, the human costs may be forgotten, much as maintenance expenses may be overlooked in the development of many types of infrastructure, from highways to bridges. Literary works have proved very successful in depicting these initial and ongoing costs and how often they are underestimated or ignored.

Administrators of the resettlement process may lack both expertise and sympathy and be insufficiently cognizant of the fact that the resettlement large dams necessitate is not truly complete once the actual move takes place. The area to which relocated populations are moved may already be inhabited, and the difficulties of integrating the old and new populations need to be considered; the newcomers may not be welcomed but might be resented or condemned, and traditional cultural ways may be inhibited. If the area to which the relocated are moved is uninhabited, that may be precisely because it is agriculturally or otherwise inferior to their original land, which may, in fact, be highly productive bottomlands now submerged by huge stagnant reservoirs. As has been observed in this context, the reality "that such [uninhabited] areas exist usually attests to their barrenness and inhospitability."[7] During the transition period, those relocated need to be properly fed and healthcare provided; this too is frequently overlooked. Rubin and Warren conclude: "Resettlement should be regarded as an integrated part of a multi-purpose river basin development project. This means that all aspects of resettlement should be directed with development in mind. They should be seen as integral parts of a long-term project, and not as piecemeal emergency measures."[8] All of this may seem painfully obvious, but experience has shown that it is by no means clear or even acknowledged. Numerous literary works attest to the validity of Rubin and Warren's concerns.

In recent decades, many environmental experts and activists have given extensive and growing attention to the consequences of failures in the areas outlined by Rubin and Warren. Thayer Scudder issued a categorical condemnation: "The evidence . . . is overwhelming that the construction of large dams has impoverished the large majority of those resettled."[9] Patrick McCully elaborates: "The great majority of those displaced by dams have statistically disappeared, swallowed up by the slums and the camps of migrant labourers."[10] Goldsmith and Hildyard reached similarly harsh conclusions: "There is scarcely a [relocation] scheme in existence that has avoided the twin problems of cultural disruption and social alienation."[11] Bruce Rich provides a long-term historical context for such problems: "Much of the history of Western economic development has also been the history of the production of huge masses of superfluous people—the creation of a new class of poor, uprooted from every traditional link to the land and the local community. The market-driven rationalization and administration of the earth's surface and natural resources for economic production has had a brutal corollary: the uprooting and depossession of huge rural populations from their less efficient modes of production."[12]

In other words, impoverished rural dwellers affected by development often become more destitute. The poor get poorer, and meaningful agency becomes more difficult.

Several authors have outlined the problems that follow relocation, many of which stem from difficulties arising prior to relocation that exert a much lengthier and deeper impact than expected. McCully, for example, discusses the rise in illness and mortality that often takes place in communities after a move and the adverse social and financial impacts on women and the elderly especially; as mentioned, dam construction is most likely to benefit young men. In part because of inadequate financial and land compensation, personal indebtedness may increase and breakdowns in family relationships and social networks occur, as those relocated need to seek employment at a great distance not only from their old but from their new homes.[13] Mark Everard points to the complications associated with what he terms *livelihood displacement*, that is, diminished economic opportunities indirectly related to actual physical relocation and often unanticipated.[14] Scudder devotes particular attention to an analysis of the multidimensional aspects of the stress pursuant to relocation: physiological, that is, health-related stress; psychological stress, which may encompass grief over what has been lost and anxiety about the future; and socio-cultural stress, which undermines community strength.[15] The cumulative picture that emerges from such analyses is gloomy and discouraging. Sadly, fictional works produced worldwide bear out the validity of the concerns about those forced to move expressed by many educated critics, academics, and environmental activists and by the increasing number of dispossessed who are activists who seek to prevent or remedy the situation caused by large dams, often with little or no effect. Literary works reveal the actuality of the many problems described above in painful detail.

Valentin Rasputin and Social Decline on the Angara

The Russian writer who most forcefully called into question the loudly touted icons of progress large dams became in the early Soviet Union was Valentin Rasputin (1937–2015). Rasputin grew up in a village on the middle reaches of the Siberian Angara that was later relocated because of construction of the Bratsk Dam, discussed in chapter 2 as a model of Soviet economic

76   DISPLACEMENT AND ALIENATION OF PEOPLES WORLDWIDE

development. Proud of his Siberian heritage, Rasputin was a vocal exponent of the loosely organized Soviet Russian literary movement that came to be known as *derevenskaia proza* (village or country prose).[16] The *derevenshchiki* embraced what they viewed as traditional rural values and bemoaned the breakdown of rural society that followed in the wake of collectivization and industrialization. In the case of Rasputin and some of the other *derevensh-chiki*, the idealization of rural Russian life was unfortunately increasingly associated with xenophobia and anti-Semitism, a tendency that became especially pronounced after the breakup of the Soviet Union. Three of Rasputin's many Siberian writings—*"Vniz i verkh po techeniiu"* (Downstream and Upstream, 1972), *"Proshchanie s Materoi"* (Farewell to Matyora, 1976), and *"Pozhar"* (The Fire, 1985)—deal with different moments in the relocation of Siberian villagers the construction of Bratsk Dam necessitated.

"Downstream and Upstream" describes a visit to his relocated hometown by the now urban-dwelling young writer Viktor. What Viktor encounters is a disorienting and to a great extent disoriented community. His family's old hut has been moved to the new settlement, but in its current setting Viktor does not recognize it; when he does, the structure appears curiously diminished and evokes a mixture of emotions: "It seemed to him now completely small and old . . . But that was it, and a warm and bitter feeling of gratitude and guilt flowed over Viktor."[17] Drinking tea and vodka with several members of his extended family, Viktor listens to their reminiscences of their former life and impressions of their new existence. For some of them, there is a marked and unpleasant contrast between a successful and rewarding past and a financially and socially diminished present. Formerly, they lived "in poverty, but happily and harmoniously," whereas Viktor's aunt Galina, who was once a prominent member of the collective farm establishment, at present works as a cleaner at the local school. Galina complains, "On the kolkhoz [collective farm] I was at least a person, but now . . ."[18] The implication is that Galina has lost status as a member of her community. There are other concerns. Viktor's grandmother recalls a fellow villager who could not bear the strain of moving and died within a year. She also regrets the submergence of the family graves. Other family members describe the impression made by the rising of the reservoir waters, with obvious sorrow. An ineffable sense of discomfort and alienation permeates the entire conversation, prompting the narrator, and the reader, to doubt the supposed benefits of this new world.

Farewell to Matyora focuses at much greater length on an earlier moment in the flooding by dam waters of long-standing settlements on the Angara. Matyora is the name of both an island and a village on the island, a village that has been in existence for 300 years. The novel describes the summer before the island will be flooded. Most of the inhabitants have moved to the village's new riverside site. Those who remain are, as in several other works discussed in this chapter, largely elderly women who are having great difficulty coming to terms with the destruction of their Edenic home and are reluctant to move to an unknown and potentially unwelcoming future: "In the evenings . . . such grace set in all around, such calm and peace . . . everything seemed so stable and eternal that one didn't believe in anything—not the removal, not the flooding, and not the parting."[19] The name of the island calls to mind the Russian word for mother (*mat'*), and the narrator suggests that the name may be related to the presence on the island of everything needed for agricultural, aesthetic, and social satisfaction. Why destroy all this bliss is the implicit question. Why abuse one's mother is the symbolic implication.

The idealization of the island's former existence stands in marked contrast to the often crude, even violent, dismantling of the island's human-made structures and destruction of natural features, like large trees, now in process. The psychic discomfort of the island's elderly inhabitants has implicitly become worse because decisions about the flooding made from above have often reached the local population in the form of rumors. A particular moment of anguish for the old women occurs when they observe what they regard as near vandalism at the village cemetery, when, without prior notice, grave markers and crosses are roughly collected in preparation for burning. Throughout the novel, the possibility of moving coffins is discussed (a few have already been moved), but in the end this does not happen, to the great distress of Dar'ia—the novel's central character, an elderly woman of strength and integrity—and other elderly characters. This problem of intended thoughtful attention to human concerns but actual neglect thereof is frequently described in literary works about relocation produced worldwide.

Generational affiliation plays a large role in the reactions to the destruction of Matyora by members of Dar'ia's family. Her middle-aged son, Pavel, is not happy about the move but looks forward to a time when at least the

78 DISPLACEMENT AND ALIENATION OF PEOPLES WORLDWIDE

process will be complete. Dar'ia's grandson and Pavel's son, Andrei, in contrast, is excited by the opportunities the changes afford and regards the dam with enthusiasm and as a chance for gainful employment. As mentioned, dam construction is most likely to provide benefits to young men eager for productive and exciting employment. Dar'ia is appalled that Andrei hopes to find work at the hydroelectric plant construction site. When his grandmother expresses the opinion that people are only temporary caretakers of the land they inhabit, Andrei pompously, and in true modernist (and Soviet) fashion, replies: "Man is the tsar of nature."[20] In other words, human beings can and will do whatever they wish to the land.

Many of the observations made in the novel about the new settlement invite immediate pessimism. Pavel mentions to Dar'ia that the cellar of his new home is filled with water. When his mother expresses surprise at such poor new construction, Pavel explains that "some stranger built it."[21] The implication is that whoever was responsible did not care about the results. This is also a frequent pattern in literary works about relocation as a result of mega-dams. Further details that emerge later about the settlement's ill-chosen location and design confirm this implication:

> [Pavel] didn't understand why it was necessary to move to this settlement, where, although it had been built beautifully and richly, the houses were right next to each other, line by line, and placed so weirdly and awkwardly that you just had to throw up your hands. And when the guys got together . . . and tried to guess why, for what possible reason it was necessary to put the settlement several kilometers from the shore of the sea that would spread out and to set it on clay and rocks, on the north side of the hill, even the silliest answer didn't come to mind . . . The explanation was simple: they [the strangers] weren't building it for themselves, they were only looking out for where it was easiest to build and were thinking least of all about whether it would be convenient to live there.[22]

Pavel also ruminates about the difficulties in finding suitable positions in the new setting for former collective farm dignitaries. As in the case of Galina, dam construction often means a loss of status and sense of social worth for the middle-aged.

Farewell to Matyora ends on a highly ambiguous note. There are still a few elderly villagers on the island, and a fog is preventing Pavel and a couple of

Displacement and Alienation of Peoples Worldwide   79

other men from reaching the island to move them to the new settlement. Symbolically, the fog evokes the atmosphere of uncertainty and disorientation that has marked the entire novel. Rasputin suggests that a farewell to Matyora is much more despair-inducing than a simple cause for nostalgia.

"The Fire" takes place in a twenty-four–hour period but includes reminiscences spanning decades by the main character, Ivan Petrovich, who, like Dar'ia in *Farewell to Matyora*, is an older person of great integrity and sense of tradition. Ivan Petrovich is symbolically linked to his home village of Egorovka, submerged twenty years earlier, by his surname, Egorov. He misses Egorovka intensely and often casts his eyes toward the space on the reservoir beneath which the former village lies. Now Ivan Petrovich lives in Sosnovka, where the chief regional industry is logging. He harbors negative feelings about the sanctioned destruction the timber industry represents: "Cutting down a forest is not the same as sowing grain, when the same tasks and concerns are repeated from season to season . . . Once you've hauled out the timber, it will be decades before there's new forest."[23] Implicitly, the rapacious timber industry undermines a sense of stewardship of the land. Deforestation leads not only to a literal erosion of the land but to an implicit attack on a sense of community.

Like the new village in *Farewell to Matyora*, Sosnovka is a sorry place: "Comfortless and untidy, the settlement was neither urban nor rural, but camplike, as if the inhabitants were roaming from place to place, had stopped to wait out bad weather and get a rest, but had gotten stuck there. But they were stuck in anticipation of when the next command to move on would come."[24] The wretched appearance of the settlement, which does not resemble a true established home, has accompanied a decline in a sense of personal responsibility, a moral collapse exacerbated by the influx of "ruffians"— itinerant laborers who have no ties to the area. Not only has the stability of the physical community descended into ruin but also that of its population.

"The Fire" of the story's title takes place at the settlement's warehouses. What should be a community effort to stem the blaze and its destruction instead devolves into massive looting, a logical outcome of the deterioration of social values that has occurred in Sosnovka. The epigraph to "The Fire" comes from a folksong: *"Gorit selo, gorit rodnoe . . . / Gorit vsia rodina moia"* (The village is burning, what is familiar is burning . . . My entire homeland is burning).[25] The physical fire described in Rasputin's story implicitly accompanies a spiritual conflagration of moral values.

80   DISPLACEMENT AND ALIENATION OF PEOPLES WORLDWIDE

Considered together, "Downstream and Upstream," *Farewell to Matyora*, and "The Fire" represent a strong condemnation of what the Bratsk and other dams on the Angara have meant for rural existence near the river. Simple comfort has given way to shoddy discomfort, and a true sense of community has yielded to rootless selfishness. Instead of an industrialized paradise, Rasputin bears witness to a dystopian hell. This is consistent with the writer's extreme nostalgia for the rural past.

Craig Lesley and the Drowning of Celilo Falls

When Euro-American explorers and later settler colonialists arrived in the Columbia River Basin, multiple Native Americans had already inhabited and profited from the area for thousands of years. The most important natural resource the Columbia long had to offer was fish, especially salmon, a plentiful and rich source of nutrition. Because of salmon's anadromous habits, a complex and heavily ritualized annual pattern of fishing, drying, and otherwise preserving it had developed at many sites up and down the river. The Native Americans who depended on the salmon generally did not live beside the river the entire year but returned regularly to the same fishing spots. In 1855 the United States government signed treaties with several groups, ensuring them certain fishing rights. The treaty signed with the Nez Perce, for example, states: "The exclusive right of taking fish in all the streams where running through or bordering said reservations is further secured to said Indians, as also the right of taking fish at all usual and accustomed places in common with citizens of the territory and of erecting temporary buildings for curing."[26] In 1904 the United States Supreme Court upheld the legality of this commitment, although erosion of Indian fishing and access to desirable fishing sites continued, in part because of competition with white fisherfolk seeking to make use of the same places. Large-scale commercial canneries also had a negative impact.

The potential complications that would arise in the twentieth century because of the imprecision and ambiguity of the phrase "all usual and accustomed places" and the construction of large dams on the Columbia could scarcely have been anticipated, although the impact on Native fishing practices and income could have been foreseen.[27] Grand Coulee, for example, was too high a dam for fish ladders; hence, there was no way for the fish to

Displacement and Alienation of Peoples Worldwide   81

continue upstream beyond the dam. This reality was conveniently ignored, with dire consequences. The construction of Grand Coulee and its reservoir, Lake Roosevelt, also meant the loss of lands by the Spokane and Colville Indians and, in regard to fishing, the flooding of Kettle Falls, a rich source of fish.

When planning for the Bonneville Dam began in the 1930s, the US government committed to identifying 400 acres of "in-lieu" (replacement) sites for "usual and accustomed places" that would be flooded, places many members of various Native American peoples inhabited for at least a large portion of the year. The government upheld that commitment for decades, but only minimally. In later years the government more often sought to settle the issue of "usual and accustomed places" by providing financial compensation rather than in-lieu sites. Embarrassingly modest financial compensation was a quick and dirty apparent solution to an ongoing problem that was creating social turmoil and increasing poverty. Nonetheless, the Pacific Northwest Indians continue to insist on their treaty-inspired rights and their right to gain at least some compensation for their losses.

Roberta Ulrich calls Celilo Falls, which was located about 322 kilometers up the Columbia River, "the largest and most significant of the Indians' traditional fishing places."[28] The site was not only an important fishing spot but was also for centuries a central location in Native American trade. The destruction of the falls occurred because of deliberate flooding produced by the Dalles Dam, located a few miles downstream from Celilo, and thus constituted a reservoir-driven immersion of the falls. Discussed, resisted, and argued about for years following World War II, the Dalles Dam was finally completed in 1957. Ulrich describes the completion scene in catastrophic terms: "Weeping Indians stood on the shore and watched as the white man's dam drowned their homes, their livelihood and the center of their culture."[29] Katrine Barber suggests that the story of the Dalles Dam and Celilo Falls constitutes a paradigm for a repeated pattern of events: "The history of The Dalles Dam illuminates the transformation of Indian-owned resources (salmon) and space (Celilo Falls) into primarily non-Native owned resources (hydroelectricity and transportation) and space (the dam and Celilo Lake [the reservoir]) through the building of a large federal work."[30]

Several Native American peoples, including the Umatilla, Nez Perce, Yakama, and Warm Springs, had long and famously fished at Celilo Falls.

82   DISPLACEMENT AND ALIENATION OF PEOPLES WORLDWIDE

Operations were conducted from wooden platforms set up on rocky out-crops over the noisy and impressive falls. The fisherfolk generally used dip nets and had ropes tied around their waists to assist in rescue efforts should they fall. It was a spectacular and often photographed sight that frequently attracted tourists. It was also a tremendous economic source for Native Americans because of the large number of salmon caught there.

The fisherfolk and their families who lived near the falls at Celilo Village inhabited an impoverished subsistence community that many non-Indians considered an eyesore. Maintenance of Celilo Village was the responsibility of several reservation governments. The Bureau of Indian Affairs was also involved in oversight of the community. Even before construction began on the Dalles Dam, one of the many dams built on the Columbia River and the one that would flood Celilo Falls, there were attempts to plan a relocation of Celilo Village and the creation of a new community, New Celilo. The so-called Celilo Indians (not an actual tribe) objected to both the intended, less convenient location of New Celilo (it would be farther from the water) and the design of some of its fishing-related facilities—facilities whose construc-tion supposed experts conceived in a thoughtless manner. Construction of the Dalles Dam, though, meant that relocation to New Celilo or elsewhere in the area became a necessity. The process was marred by ignorance, confu-sion, and, among some whites, a complete lack of sympathy for the Native American losses suffered. At the time and later, disputes over Native versus non-Native fishing rights, the rights of certain Native American peoples ver-sus others, and federal versus tribal jurisdiction exacerbated the situation. In other words, judicial chaos ensued and worsened matters, although Native Americans were sometimes able to maintain their treaty-granted rights.[31]

Craig Lesley (b. 1945) is an Oregon native with a keen interest in environ-mental destruction and its impact on Native Americans in the area. Winterkill (1984) and its sequel, River Song (1989), set in the 1980s, treat in partially ret-rospective fashion the experience of the Nez Perce and other Pacific Coast Native Americans in the nineteenth and twentieth centuries. The major pro-tagonist of both novels is the Nez Perce Danny Kachiah, who at age thirty-four is becoming too old to pursue the itinerant life of a rodeo cowboy—a life that epitomizes the haphazardness and insecurity that now characterizes much Native American existence. Danny is the divorced father of teenager Jack, whom he brings home to live with him early in the novel after the death

Displacement and Alienation of Peoples Worldwide   83

of his ex-wife. The novels are based on meticulous historical research. Lesley received several awards for *Winterkill*.

Celilo Village appears at the beginning of *Winterkill* when Danny stays overnight in a corner of its longhouse. The initial description of the site evokes an immediate feeling of loss and destruction: "The old village with its salmon-drying shacks and Wy-Am longhouse was gone now that the dam's backwaters had covered Celilo Falls and ended the fishing."[32] When Danny closes his eyes, he imagines he can still hear the roaring of the falls, and he regrets that his son will never see them.

After returning to his trailer on the Umatilla Reservation near Pendleton, Oregon [the Umatilla are one of several tribes that made treaties with the white settler colonialists in the 1850s], Danny thinks back to a visit to Celilo he made as a child with his now deceased father, Red Shirt, before the falls were flooded. The traditional process of fishing for salmon is described in loving detail. Danny and Red Shirt attend a celebration with Red Shirt's friend Sammy Colwash, a resident of Celilo Village who is anxiously anticipating the closing of the Dalles Dam. "We still have our fishing rights, but what good are they if the dam swallows the falls? The fish are more than money," states Sammy.[33] Fishing at Celilo is a way of life, not simply a source of income, he implies.

Danny and his father return to Celilo Falls the day the floodgates are closed. Together with Sammy and numerous spectators, they watch the dam waters creep toward the falls. Afterward, Sammy moves from Celilo and raises sheep: "The dam waters had swallowed the falls, and Sammy said it made him sick to look at the new lake, so he'd bought some sheep and moved up the Umatilla."[34] After Sammy dies, Danny asks his father what Sammy died from. Red Shirt responds, "He was never the same after moving from Celilo."[35] Red Shirt's simple observation conveys a sense of Sammy's psychic disorientation after the loss of Celilo Falls.

Later in the novel, Danny attempts to explain the significance of Celilo Falls to his son. In response to a comment by Jack, Danny concedes that while the fishing at Celilo Falls was not highly profitable, it did provide a living and a sense of community. The money the government paid the Indians for the loss of fishing rights was quickly spent, says Danny—a common pattern when government funds are used to recompense Native Americans for their losses—and the Celilo residents' harmonious existence disappeared:

84   DISPLACEMENT AND ALIENATION OF PEOPLES WORLDWIDE

"After the dam, there were no more fish. Before that, no matter what happened, the people always had the fish. For the Celilos, fishing was a way of life."[36] Danny describes how on the day the floodgates were opened, the old people turned their backs on the rising water and pronounced a death lament for the falls. As for the present, Danny observes: "If you look into the old people's eyes, you can tell they are still dreaming of the falls and the salmon."[37] The poignant loss the diverse Native Americans at Celilo suffered serves as an especially painful example of the multiple losses sustained by the Native peoples of the Pacific Northwest. Once again, it is the elderly who suffer most.

River Song revisits and adds further details to Danny's memories of Celilo Falls, mentioning, for example, ancient petroglyphs the dam waters flooded or that in some instances were ripped from the canyon walls by museum collectors. Much of the novel's narrative, though, focuses on Danny's increased involvement in fishing at the Wind River in-lieu site. There, an old Celilo inhabitant Willis Salwish introduces Danny and Jack to the intricacies of contemporary fishing on the Columbia River, and Danny discovers there may be a more traditional and satisfying future available to him through his own actions than he had anticipated. Willis's family is one that has resisted the government prohibition against permanent residences at the Wind River site by the Bureau of Indian Affairs. Their resistance has ideological underpinnings: "By living on the sites, Willis believed they were exercising ancestral claims and preventing the land from being taken away again, perhaps sold to make parks catering to windsurfers or expanded facilities for the white sportfishermen."[38] Willis's fears point to the threat of ongoing economic and cultural displacement that affects the Pacific Northwest Native peoples and to the eternal problem of the losses of the disadvantaged few for the often frivolous gains of the affluent many.

When Danny and Jack arrive at the Wind River site, Willis's grandson, Jimmy, has just drowned because of a nighttime collision between Willis's boat and a state fisheries boat on the lookout for illegal fishing. Willis is increasingly bitter, declaring: "When they get us all pushed off the river, maybe they can build more places for the tourists and windsurfers. Maybe they can put up a nice little museum here with statues and pictures, so the gawkers can see what Indians once were like. They'll have to carve wooden salmon, too, because the Creator said when the old-timers quit catching His

Displacement and Alienation of Peoples Worldwide 85

fish, He'll quit sending them."[39] This points to the actual massive decline in fish population that occurred after the creation of multiple dams on the Columbia River, a true financial loss. According to Native custom, because of his grandson's death, Willis cannot fish for a year. Danny offers to fish in Willis's stead. For Danny, fishing on the Columbia is an important step in his simultaneous return to traditional practices and his spiritual and emotional maturation, what J. C. Davies terms "the changing, learning center of a *bildungsroman.*"[40] This step represents the beginning of the process of Danny becoming an effective agent, not a victim.

Danny and Jack's venture into fishing exposes them to anonymous acts of anti-Indian vandalism, in the form of slashed nets and arson at the fishing camp. In speaking to the investigating police officers, Willis's granddaughter, Velrae, is outspoken in her denunciation of the crimes: "This is an act of war against a sovereign nation! Someone has declared war on the River People!"[41] Velrae, an un-stifled activist, gives voice here to the strong sense of persecution and assault on Native American fishing rights that constitutes a major theme of *River Song* and the need for resistance. The vandalism does not deter Danny, and he begins to consider buying his own boat and becoming Willis's fishing partner. He is increasingly drawn to traditional fishing ways. At the Wind River Narrows, he engages with pleasure in fishing with a net from a wooden platform. His former fear of drowning while fishing in this way has gone: "Now he was no longer afraid. Although these Wind River Narrows weren't Celilo, they gave him a sense of traditional fishing, and he understood why his ancestors had come to the river. It made him sorry that Red Shirt had relinquished his fishing sites, and he vowed he would try to coax Jack onto the platforms next year."[42] At the end of *River Song*, Danny has an almost mystical vision of salmon on the Columbia and other rivers of the Northwest: "Danny imagined all the waters pouring toward the Columbia. Clearwater and Snake, Deschutes and White, Klickitat and Wind . . . The salmon rode the swelling currents out to sea, then beat their flesh against water and rock, returning. Always returning."[43] This vision stands in marked contrast to the reality of the salmon-poor status of the Columbia River, but it provides hope for a future richer in fish and possibilities for traditional Indian life.

Danny's commitment to fishing on the Columbia River echoes the attitudes expressed by his great-grandfather Left Hand in the late nineteenth

86 DISPLACEMENT AND ALIENATION OF PEOPLES WORLDWIDE

century. Left Hand had been a Dreamer Warrior, a member of a Nez Perce faction that refused to accept Christianity and secretly performed ancient ceremonies. The Nez Perce and other Pacific Northwest Indians have experienced multiple losses, and the drowning of Celilo Falls is the central loss highlighted in Lesley's novels. At the same time, though, the Nez Perce have engaged in creative acts of resistance against the depredations of US settler society. Like Left Hand, Danny refuses to abandon traditional ways completely and in the process becomes a stronger and more mature individual.

Sioux Losses on the Missouri River

Native Americans have been moved for centuries, by force, by treaty, by trickery. Often told that they could continue west, the west in question turned out not to be far enough west for the settler colonists who followed the Indians. They became convinced that if they became "civilized" they could stay in place, but that wasn't true either. Many Native Americans were forced to move, for example, to what is now Oklahoma; the Cherokee are the best-known case. Only in the twentieth century, however, were certain peoples forced to leave their fertile riverside homes because of the construction of mega-dams. This represents but one instance over a long period of forced removal of Native Americans.

The Missouri River was notorious for its periodic and at times dangerous flooding. It was also long eyed by engineers and lawmakers alike because of the possibilities it afforded for irrigation and the generation of hydroelectric power. In 1944 the Pick-Sloan Missouri Basin Program, a compromise plan for development authored by Lewis A. Pick of the Army Corps of Engineers and William Glenn Sloan of the Bureau of Reclamation, gained approval. The plan included the construction of several large dams, some of which were located near Sioux Indian reservations on the Missouri. The historian Michael L. Lawson forcefully condemns and outlines in detail the negative impact of the Pick-Sloan program on the Sioux: "The Pick-Sloan program has demonstrated the sham of native rights and the hypocrisy of federal Indian policies in the twentieth century."[44] No one consulted the Native Americans before construction began. The issue of compensation was poorly considered. The Native American tribes involved were put in the position of having to accept virtually whatever the United States Congress approved. Lawson is not alone in his criticism of Pick-Sloan. The historian Donald L. Parman asserts that while the percentage of Indian land flooded on the Missouri was comparatively small, it contained most of the areas used by local Native Americans for subsistence: "Only 6 percent of the Sioux's land, for example, was flooded, but approximately one-third of the residents underwent relocation. In addition, resettlement disrupted long-standing communal and kinship ties and deeply felt attachments to the land."[45] Once again, we see a familiar pattern of dislocation and alienation.

Five Sioux reservations in North and South Dakota were particularly affected by the building of the Oahe, Fort Randall, and Big Bend Dams: Standing Rock, Cheyenne River, Yankton, Crow Creek, and Lower Brule.[46] Hundreds of thousands of acres were flooded, and hundreds of families were "forced to move from rich, sheltered bottomlands to empty prairies."[47] The earthen Oahe Dam, constructed near the state capital Pierre, South Dakota, created one of the largest human-made reservoirs in the United States. The economic life of the Sioux—already precarious—suffered, as did social and religious life. Relocation and the reconstruction process were poorly managed and marked by federal paternalism and a degree of chaos and poor planning; some families had to move twice. Unemployment was widespread. Fewer wood, game, and other natural food sources, such as wild fruit, were available. Federal and tribal services were disrupted. Although in some instances living conditions did improve, for example, through the provision of electricity and indoor plumbing, the actual use of such improvements by Native Americans was often hampered or prevented by problems of affordability. Hence historians' grim conclusions about the ultimate impact of the Pick-Sloan program on the Sioux.

The historical assessment of Pick-Sloan has been confirmed by Elizabeth Cook-Lynn (b. 1930), a Crow Creek writer and the first female author discussed in this study, who has long expressed strongly critical views of the relations between Native Americans and the government. Cook-Lynn has received multiple awards for the effectiveness of her writings. The social and economic devastation wrought by dams on the Missouri permeates the moving narrative of her *Aurelia: A Crow Creek Trilogy* (1999).

The three short novels that constitute the *Aurelia* trilogy—*From the River's Edge, Circle of Dancers*, and *In the Presence of River Gods*—focus on several periods in the life of Aurelia Blue, born in 1933 and a lifetime inhabitant of the

88   DISPLACEMENT AND ALIENATION OF PEOPLES WORLDWIDE

Crow Creek Reservation near Pierre, South Dakota. *From the River's Edge* is set in the mid-1960s, *Circle of Dancers* in the early 1970s, and *In the Presence of River Gods* in the early 1980s, with an epilogue from 1995. Throughout the three novels are constant reminiscences of and references to events from the 1950s.

From the River's Edge centers on the young Aurelia's long relationship with an older married man, John Tatekeya. The beginning of their relationship is linked to the despair John feels at the time of the building of the Oahe Dam, when "he had been forced to move his cattle, his home, and his outbuildings out of the way of the backwaters of the hydropower dam called Oahe."[48] The homes, cemeteries, and churches of John and many of his family members and neighbors are relocated at the last minute, and none of the many promised benefits materialize: "For years after that he had none of the promised electricity, no fences to hold in his cattle and horses, no water except the ever-seeping backwaters, no barns for storage, and no haystacks to feed his stock for the coming winters. It was a time when his wife nearly gave up, a time of great stress, waste, and confusion."[49] John embarks on his affair with Aurelia in the late 1950s, in part in an effort to ease the pain he feels at this tremendous loss. In other words, adultery is symbolically an outcome of alienation.

In John's eyes, the fault for this unhappy turn of events for the Crow Creek people lies explicitly with the white world and its abiding pernicious attitudes toward nature and Indigenous peoples. In 1967, John is a participant in a court case involving the theft of some of his cattle. In a conversation with his white lawyer, who suggests that his client has too much pride, John thinks to himself: "How can you chastise me, a Dakotah, about arrogance when it has been your people who have forced your religion on everyone throughout the world, your people who changed the rivers that we live by and flooded our lands?"[50] In John's mind, the breakdown in social relations that has contributed to the theft of his cattle is one of many socially unfortunate consequences of the upheaval caused by the dam. Moreover, the construction of the dam itself is just one more event in a long series of actions by white people in the service of "the exploitative and brutal policy of colonialism."[51] Cook-Lynn points here to a key negative result of settler colonialism and, at the same time, to continued Native American resistance to such practices.

Displacement and Alienation of Peoples Worldwide 89

Aurelia interprets the significance of the dam in much the same way as does John. She too links social breakdown with the impact of the dam on the Crow Creeks. Like John, Aurelia views their relationship in part as a result of the dam: "She knew now that the flooding of the homelands had to be taken into account in any explanation of her devotion to [John] Tatekeya, and his to her."[52] More broadly, Aurelia regards violence done to the earth, with the dam an example, as inseparable from violence done to peoples by other peoples, to settler colonialism writ large. In a sense, the end of Aurelia's relationship with John that comes about in the course of *From the River's Edge* is, like the beginning of their relationship, symbolically linked to the social, economic, and emotional havoc caused by the dam; the dam has destroyed not only a desirable physical permanence but continuity of emotional commitments as well.

The second novel in the trilogy, *Circle of Dancers*, deals with Aurelia's relationship with a younger man, Jason Big Pipe, and takes place against the background of the development of the American Indian Movement (AIM) in the early 1970s—a key example of Native American activism. Once again, the damming of the Missouri haunts the memories of the novel's characters. We learn that Aurelia's maternal grandfather, Amos Two Heart, was complicit in the agreement made regarding the Sioux and the dams. Educated and well-to-do, Two Heart played a formal leadership role in the Crow Creek's dealings with the United States government. Considered a collaborator by some, an influential figure by others, Two Heart is denounced by Aurelia's paternal grandmother: "On one of his trips to Washington, D.D. [Two Heart] signed the documents permitting the flooding of hundreds of thousands of acres of tribal lands, for the federal government's hydropower development of the Missouri River . . . The Mni Sousa [Sioux name for the Missouri] was no more."[53] Aurelia's grandmother's negative assessment of Pick-Sloan was widely shared by other Native Americans: "Everyone, after a few years, said that the destruction of the mythic river simply brought on more 'relocations,' more 'terminations,' more so-called anti-poverty schemes."[54] Two Heart's involvement in this process points to the corruption, co-optation, and disingenuousness that are often part of large dam construction schemes. The AIM constituted rejection of a position of passive victimhood.

The narrator of *Circle of Dancers* issues additional condemnations of government actions linked to Pick-Sloan. Discussing a farm and ranch program intended to revitalize subsistence ranching of cattle and pigs, the

90   DISPLACEMENT AND ALIENATION OF PEOPLES WORLDWIDE

narrator comments: "This plan, the politicians told everyone, would cost the American taxpayer nothing, since the thousands of acres of flooded Indian lands, the miles and miles of destroyed timberlands and river bottoms had already paid the sacrificial price for such 'progress.' They seldom publicized the fact that Indian-held lands had been reduced by five hundred and fifty square miles nor that each Indian had received a $400 check for his loss."[55] Sioux characters in the novel involved in the farm and ranch scheme discover that it does not yield the promised benefits and only serves to lead them into deeper poverty. Others are driven into wretched existences off the reservation. "Price out Indians and ship them off to live in the Twin Cities, the ghettos of L.A., on some federal subsistence grant money," thinks Aurelia.[56] Once again, dam construction has led to destruction of rich lands, laughable compensation, and an uprooted society.

Much of the background narrative of the third novel in Cook-Lynn's trilogy focuses on the 1980 kidnapping, rape, and murder of a young Crow Creek girl by white men, who are finally brought to justice in 1995 in a victory of sorts for her relatives and friends. In the context of *In the Presence of River Gods*, this case implicitly reflects widespread predatory and racist attitudes whites hold toward Native Americans. Toward the end of the novel, the narrator makes the implicit explicit: "No one except Indians connected the atrocity to the matters of racism, to the long simmering hatred of one race toward another, to the matters of destination and ambition and journey's end."[57] Racism, it appears, is a natural and inevitable companion to settler colonialism, but, as the novel shows, it can be fought.

As in *From the River's Edge* and *Circle of Dancers*, repeated references to the dams on the Missouri in *In the Presence of River Gods* underscore the prevalence and persistence of such attitudes. During the novel, Aurelia and her children stay temporarily with her old friend Connie, who lives in a reservation town built under the auspices of the Bureau of Indian Affairs. The impression the community makes on Aurelia is grim:

> The community where she stayed now with friend Connie seemed to be a
> place newly carved out of a vast, flat prairieland. In fact, it was just that. This
> Agency town, like others up and down the Missouri River, had been what
> they called "relocated" when huge hydropower dams were built upstream
> and downstream. The wind roared through its streets in the winter; horribly

battered men, some of them quite young, hung about the street corners and the cafe in good weather and bad. Beer cans and wine bottles were thrown into the weeds along the walks. In the recently designated park in the center of town, men would be found lying as if dead, the telltale white rim about the mouth. Huffers with dead eyes and agonized grimaces.[58]

After the young Indian girl's body is found, Aurelia links the murder to the damming of the Missouri and whites' destructive attitudes toward Indians: "Aurelia remembered Tatekeya's grief concerning the destruction of the river and she felt that the white man's treatment of the earth, always suspect to herself and others, was one of the terrible things that reverberated throughout all of the human relationships surrounding them. Now, for Aurelia, at least, the actions toward both the earth and humans were no longer shrouded in ambiguity. The destruction was all around."[59] The environmental and social consequences of US machinations with the Missouri have been huge and profoundly detrimental.

A Crow Creek Trilogy is by no means entirely lacking in optimism and a sense of the possibilities for agency. Aurelia and many of the other Sioux characters are strong individuals who do not succumb to the injustices that have been done to their people. References to ongoing Indian attempts to reverse the depredations white Americans have carried out since the nineteenth century, such as acts by the AIM, point to the potential for serious restitution. The overall tenor of the three novels, however, is bleak. The economic, social, and cultural devastation visited upon the Crow Creek Sioux as a result of the Missouri dams is only one instance of ongoing devastation. It symbolizes the parallel destruction of peoples and the earth and points toward both environmental and social upheaval.

Aswan and the Nubians

As mentioned, the number of people affected by mega-dams gradually grew throughout the world. Egypt's Aswan High Dam is a case in point. The construction of dams, both low and high, on the Nile near the city of Aswan had been considered by engineers since the late nineteenth century. The first Aswan Dam was initially completed in 1902 and subsequently raised twice, first by five meters and later by nine meters. The second, so-called Aswan

High Dam was originally the brainchild of the Greek Egyptian engineer Adrian Daninos, who first argued in 1912 for a large hydroelectric dam to assist with industrial development. It was only after Gamal Abdel Nasser and other members of the military overthrew the Egyptian monarchy, however, that the idea received serious official attention. A high dam at Aswan afforded the new government precisely the kind of symbol it sought in implementing its vision for a modern Egypt.[60] Most important, such a dam "would free Egypt from being the hostage of upstream riparian states by providing for century storage within the boundaries of Egypt."[61]

Several years passed before construction on the Aswan High Dam, a few miles upstream from the old Aswan Dam, began. Initial plans had British and US support, but when that support evaporated because of political disputes (the British and Americans were suspicious of the strength of Egypt's commitment to the West in the Cold War), the Egyptians turned for assistance to the Soviet government, which was eager to help them and thus gain greater influence in the Middle East (figure 3.1). For Nasser and his associates, the dam was no longer "simply a big engineering project, but . . . the symbol of Egypt's will to resist imperialist [read US and British] endeavors to destroy the revolution."[62] In 1959 Sudan and Egypt agreed on conditions for the construction of the Aswan High Dam; Sudanese agreement was necessary in part because the reservoir that would be created would reach more than 160 kilometers into Sudan. The construction would displace more than 100,000 Nubians, half of them Sudanese. The Nubians had already been affected by the construction and later raising of the first Aswan Dam. Because of the construction of the Aswan High Dam, the Sudanese Nubians had to move from the Wadi Halfa District to distant Khashm al-Girba near the Atbara River, and the Egyptian Nubians needed to move to recently reclaimed lands in the Kom Ombo Valley, north of the city of Aswan. The Egyptian resettlement area is now called New Nubia; the Sudanese area is New Halfa. The completion of the dam was officially celebrated in 1971.[63]

Nubia has served as a link between Egypt and sub-Saharan Africa for thousands of years, and the Nubians have been a source of Egyptian enslavement and later an object of settler colonialism. The area that was flooded was home to remarkable ruins from various cultures, interest in whose salvage was generated internationally at the time of the Aswan High Dam's construction. Ironically and sadly, it has been argued that in fact "there was greater interest

FIGURE 3.1. *Gamal Abdul Nasser with Nikita Kruschev, during the Ceremony of the Divert of the Nile at Aswan High Dam*, photograph, 1964. Wikimedia Commons, https://commons.wikimedia.org/wiki/File:Gamal-002.jpg.

in history than in people."[64] Recall the similar comment made in chapter 2. Ultimately, more than forty countries took part in salvage efforts, the most famous of which was the raising of the rock-cut temples built under the Egyptian pharoah Ramses II near the village of Abu Simbel. The great temples and monuments were raised, in an impressive engineering feat, and now stand on the western bank of the reservoir known as Lake Nasser.

The most obvious benefit stemming from the construction of the Aswan High Dam was the provision of a regular means of irrigation for agriculture and an important backup source of water during times of drought and low flood levels. The negative side effects that emerged were, as usual, unanticipated or underestimated. They include "(1) the water loss through seepage and evaporation; (2) the sedimentation of the Nile silt in the lake and its impact on land productivity ... (3) the waterlogging and level of soil salinity from year-round irrigation; and (4) the problem of disease menace through the expected increase in the incidence of schistosomiasis [a parasitic snail-borne disease] due to the extension of land irrigation."[65] What had been touted as Egypt's route to modernity often came to be popularly viewed instead as the source of serious economic challenges and other problems.[66]

The impact of the Aswan High Dam on rural Nubian existence was calamitous. Nubian society had already long exhibited a pattern of male labor migration: "Egyptian Nubian males have, for several generations at least, worked in Cairo and other large cities, leaving farming and other economic activities to their wives and to males who because of age, infirmity, or for other reasons were unable to work there or were not interested in urban careers."[67] This pattern intensified after the construction of the Aswan High Dam and contributed to what one Nubian termed "a dam complex": "According to him the first dam forced Egyptian Nubians to leave their 'beloved' land and seek work in the 'unfriendly cities' where they never felt at home; but the second dam was a total curse which ended a life-style that satisfied their basic needs and gave them peace and happiness. Labor migration had split families and created communities where females outnumbered males. The outcome was quiet Nubian villages dependent on cash remittances and patiently awaiting the return of absent men."[68] A pervasive atmosphere of social depression in New Nubia was exacerbated by loss of proximity to the Nubians' beloved Nile and resettlement into homes whose basic design was significantly different from and inferior to that of their previous dwellings and often necessitated generational fragmentation of families. The new settlements might offer better community services, provisioning, and educational opportunities; but their inhabitants also suffered from a greater incidence of stress-related diseases, such as high blood pressure.

Two Egyptian Nubian authors who have addressed the Nubian resettlement in their Arabic-language writings are Idris Ali (1940–2010) and Haggag Hassan Oddoul (b. 1944). Ali was born in Aswan. Largely self-taught, he spent much of his life in poverty in Cairo. Oddoul was born in Alexandria and worked in construction on the Aswan High Dam in the 1960s. He began writing when he was in his forties. Both Ali and Oddoul are representative of a trend in Nubian literature that has been termed "post–Aswan Dam literature," in which "the themes of loss, dislocation, and alienation are central."[69] In the case of the Nubians, the literature that has been produced offers exposure of a gross and grotesque example of settler colonialism. The Egyptians are often racist in their attitudes toward the Nubians and regard them in many ways the way some white Americans regard Native Americans, as shiftless and undeserving.

Idris Ali's novels *Dongola: A Novel of Nubia* (1998) and *Poor* (2007), especially the latter, include autobiographical elements drawn from the author's often marginalized existence. The male protagonist of *Dongola*, the first Nubian novel ever translated into English, is Awad Shalali. Awad has spent much of his young adult life in Cairo and ten years in an Egyptian prison camp because of his political activities, one of the many signs of his refusal to accept Egyptian victimization of the Nubians. His father, a former farmer, had long worked in Cairo, where he married an Egyptian woman who exploited him financially until his death. At the beginning of the novel, Awad pays a brief visit to his home village in New Nubia, where his mother, Hushia al-Nur, still lives, but he is soon forced to move on—this time across the border to Sudan, the first of many countries he will visit. Years later Awad returns after accidentally learning that his mother has been reduced to extreme poverty and is heavily dependent on the goodwill of neighbors. Although Awad now intends and is able to send his mother money regularly, Hushia wants to marry her son to a local girl so she will, in keeping with tradition, have someone who will be obliged to take care of her without compensation and thus provide her with greater security.

The Dongola of the novel's title refers to the capital of medieval Nubia, which now lies beneath the reservoir of Lake Nasser, like many monuments from antiquity. For Awad, Dongola represents the potential for the Nubian political resurrection for which he longs and strives. In Awad's eyes, his homeland has become "a water reservoir for the north."[70] The Aswan High Dam is a particularly egregious example in a long series of abuses carried out against the Nubians by Egyptians and the British: "He cursed them all, including the wanton Nubian King Shakandah, the traitor. He cursed Kanz al-Dawla and Sir William Wilcox, chief of the treasury, the pashas, the sultans, Muhammad Ali Pasha, and all the ministers of irrigation who had ever lived and those still to come, and anyone who had placed a single stone in the first and second dams. He cursed the river that had surrendered to the dam, and cursed the whole world, which had helped to save the temples, while leaving the people to their fate."[71]

Awad sees drought, famine, disease, ugly and poorly designed new homes, and male labor migration as direct consequences for the Nubians of the construction of the Aswan High Dam. At an impromptu neighborhood celebration of his first visit to his mother, gaiety turns to weeping when Abdu Shindi, a local Nubian singer, extemporizes a folksong about the Nubia that has been lost. Shindi's song captures the imagination of the relocated Nubians even after this gathering: "His sad song had spread among old and young; they all sang it and wept with sorrow, the women sitting and waiting, remembering, as their eyes teared up, and words brought back the men's memories of the streams and roots, and they too were overcome with despair and homesickness."[72] Awad fantasizes about bringing a lawsuit against the dam builders and restoring the borders of Old Nubia. He asks an Egyptian intelligence officer who is investigating him for possible subversive activity whether the Nubians are not as important as temples and statues, echoing once again thoughts about the value of antiquities over people. The mayor of Awad's mother's village rejects the notion that New Nubia is a disappointment, arguing instead that resettlement has brought the Nubians many benefits, for which they should be grateful to the Egyptian government. Implicitly, his remarks reflect a craven collusion with the authorities, the kind of corruption Aurelia's grandfather evidences in *A Crow Creek Trilogy*.

Awad's mother, Hushia, has suffered the effects of both the first and second Aswan dams for decades. Her father, an enterprising farmer, perished by sinking into deep mud caused by manipulation of the reservoir water from the first dam. Her husband left for Cairo almost immediately after Awad's birth. Her son also moved to Cairo and became disaffected and radicalized. When the resettlement following construction of the Aswan High Dam took place, her neighbors had to force Hushia to move. Her brother, on whose financial assistance she relies, died after spending all his money on a lengthy battle with an incurable disease. Finally, Awad returns for the second time. Disenchanted with his dreams of a resurrected Nubia after years of unsuccessful political activity, Awad has spent further years in Europe: "He cursed the north, the south, Nubia, Dongola, and his comrades, and departed, fleeing the fire of the sun and the whole dark continent."[73] Now Awad is in love with and wishes to marry a wealthy French woman, who, ironically, had been active in the campaign to save Abu Simbel but apparently shows little interest in the Nubians' plight. Hushia finds the idea of her son marrying a foreigner intolerable. With the help of the few men left in the village, she convinces Awad to marry locally instead. His bride is Halima, who has provided help to his mother. Awad departs almost immediately after the wedding and consummation of the marriage.

Displacement and Alienation of Peoples Worldwide   97

The third and final part of *Dongola*, "The Sorrows of Hushia and Halima," focuses on the particular desperation of the female Nubian experience: "Halima waited and her waiting grew long because she was like the other forsaken women of Nubia, all of whom were waiting for men who had journeyed far away, to the cities of Egypt, the Arab lands, and overseas. They lost sight of time; they got lost in its tracklessness and were dazzled by its passage . . . But Halima's marriage, and her wait, were unrivaled."[74] According to custom, Halima must now obey orders given to her by Hushia, who becomes increasingly decrepit, querulous, and demanding. When Halima appeals to her family for sympathy, she receives none. As the years pass, Halima's mental state becomes highly unstable. Eventually, she seduces an itinerant Upper Egyptian carpenter. Discovered *in flagrante delicto* by Hushia, Halima strangles her mother-in-law and blames the carpenter for her death. As the novel ends, a crowd of villagers is hunting down the carpenter. The social breakdown precipitated in New Nubia by the High Aswan Dam has resulted in horrific tragedy for Awad's family. It provides an excellent example of the particular harm women and the elderly often endure in the wake of dam construction.

Poor touches on many of the same themes as *Dongola*. It opens in 1994 in Cairo with the first-person narrator's intention to commit suicide because of his wretched existence. Subsequent chapters are retrospective. Imagining a request by an Egyptian woman to tell his story, the narrator thinks to himself:

> Summon your memory and tell her the story of a Nubian child of the post-Aswan generation. Tell her frankly, "People of Egypt! This is what you have done." Without fear, without embarrassment, ask them, "What have you done to this country? This country that is also your country! Why did you dam up the Nile over Nubia's lands? Why do your trains stop at Aswan? Why did you draw the border at Aswan, and with it, the government and public services and development schemes?" Tell them about their hero Mehmet Ali who cut Nubia into halves. Tell them about how the pharaohs treated Nubia. Tell them about their mythical hero Baybars who butchered the people of Nubia . . . Ask them, "Why did you take a nation of farmers and turn them into servants and doormen for your palaces and villas?" Tell them, "Building us a museum is not enough . . ." Tell them, "It's no good to talk

about working tighter in solidarity—about us making a sacrifice for the good of the greater national projects—if you're not going to pay us the compensation that we're owed."[75]

The history of Egyptian exploitation of the Nubians is indeed a lengthy one. The narrator of *Poor*, it emerges, grew up in chaotic poverty and a state of malnutrition in Nubia. His family and neighbors suffer from reservoir floods caused by the first Aswan Dam. In the wake of famine, most of the men—including the narrator's father—depart in search of remunerative labor elsewhere. Eventually, the boy decides to go to Cairo himself, although a relative warns him that without adequate education, he will embark upon a wretched existence: "Cairo is a big flesh-eating demon . . . To live there with your honor intact, you need to arm yourself with an education . . . If you go there in your present condition, you'll fall into the hell of Cairo's service jobs and come to hate the day you were born. An entire generation of Nubians struggles valiantly in night schools just to find the lowest place among the people of the North."[76] The boy does not listen and does indeed frequently encounter the Egyptians' racist condescension, which he refuses to passively accept.

Arriving in Cairo, the narrator is struck by the tremendous disparities of wealth and poverty. He finds his father living in a dilapidated garbage-strewn neighborhood with a slovenly Egyptian woman who is clearly taking financial advantage of him, a similar situation to that of Awal's family in *Dongola*. A few years later, the narrator's entire family moves to Cairo. The narrator tries to stay in school but fails, despite his voracious reading. He works a series of menial jobs and has a variety of often sordid relationships. He has violent run-ins with the police. Cairo proves to be an existential dead end, and the narrator repeatedly contemplates suicide.

Oddoul's *Nights of Musk: Stories from Old Nubia* (2005) includes four stories, two of which—"Adila, Grandmother" and "The River People"—focus on what has been lost after the construction of the Aswan High Dam. "Adila, Grandmother" begins with the burial of the grandmother of the narrator, Mohamed. Mohamed, who is of mixed Egyptian and Nubian parentage, recalls earlier visits to "the exiles' village" and interactions with his grandmother and his Aunt Awada.[77] He describes how he gradually came to appreciate his relatives and the pathos of their existence: "I mixed more with the villagers and slowly gained a deeper understanding of the barren poverty of their lives in a strange land. So far, some of the village land had been cultivated, but most was still dust. From afar, it looked like a mangy scalp with patches of hair amid its baldness."[78] This unsavory simile captures perfectly the sorry state of the new settlement. Further, it is a typical example of the poor quality of the new land provided to those relocated by dams everywhere.

Mohamed's grandmother is filled with anger over the aftermath of the move: "Why have they driven us to this arid, God-forsaken place? Where is our old village? Where is our Nile? Where are our palm trees and our spacious houses?"[79] She laments the growing need for Nubian men to engage in migrant labor and expounds upon the inferior quality of the land in the new settlement: "They've pulled us up by our roots, and we've become like brushwood. Our sons went off all over the place to work as servants in the land of plenty . . . And we here, they have thrown us into the valley of the demons. They gave us this land. Nothing grows on it but evil plants with bitter fruit that even the animals loathe. They've killed us."[80] Saddest of all, perhaps, is the situation of Mohamed's aunt, who repeatedly mentions the name Yaseen in nighttime deliriums. When Mohamed asks about Yaseen, Awada finally tells him: "We got engaged, and then when they built the dam and our lands were drowned, he went north to make his fortune and then was supposed to come back and marry me. He never sent any money, and he never came back."[81] Social dislocation, in Awada's case, leads to a bitter and lonely existence.

This theme of female Nubian abandonment, familiar from Ali's *Dongola*, informs the major plot of "The River People." The River People are folkloric beings who live in constant bliss. The protagonist and first-person narrator of large portions of the story, Asha Ashry, is a beautiful young woman in love with a neighbor, Siyam, who, like Awada's beloved, leaves "when the dam from the north fouled our valley."[82] Asha awaits his return for years, refusing to consider marrying anyone else. She loathes the new existence that has been forced upon the Nubians: "Dam piled high, you are the same age as me. You split up lovers. They dumped you in the way of the mighty river. You have blocked the life-flow of water. Behind you it has built up and drowned half our land. The river is good like its people, but the dam confined the water in a huge lake . . . It ruined the time of peace and purity.

We moved out, leaving behind our cool spacious houses for cramped sweltering ones that hung on the side of the mountain like carbuncles."[83] Like Mohamed's grandmother in "Adila, Grandmother," Asha also laments the need for Nubian men to engage in demeaning labor in the north. She is tormented by stories she hears of Siyam's involvement with a Greek woman in Alexandria. "The dam has destroyed my life," concludes Asha.[84] She wishes for the dam's destruction.

Finally, Asha receives news that Siyam is returning. He is ill, though, and drowns when the mail boat on which he is traveling sinks. Significantly, the accident takes place opposite the statue of Ramses at Abu Simbel, another ironic comment on what was considered worth saving and who was not at the time of construction of the Aswan High Dam. The catastrophe evokes mass weeping and mourning by the Nubian women: "They wept at the fate of their husbands and sons in exile. Terrified of an unknown future, they bemoaned their lot, the migration of the men to the north, to the painted white women of the north, and the danger of seduction. They were left with the burning heat of the sun and the parched earth of their drowned land."[85] Asha goes completely mad and drowns herself. Her fate is as terrible as that of Halima and Hushia in *Dongola*. In the wake of the construction of the Aswan Dam, both Oddoul and Ali suggest, the Nubian women suffer even more than the men.

Development Casualties in India

Post-independence India was obsessed with development in a manner that made Egypt look conservative. Among other preoccupations, the use of modern irrigation methods seemed to promise an agricultural boom, an important part of which came to be called worldwide the Green Revolution. Huge dams were intended to play a central role in this route to escape widespread poverty, hunger, and undesirable dependence on other countries.[86]

The construction of large hydroelectric dams figured prominently in India's first Five Year Plan (1952–57) and met with widespread enthusiasm in the mid-twentieth century. Big-dam building, especially multipurpose dams, was central to this vision of development. Once again, the people who would benefit were "what has been termed India's 'dominant coalition of proprietary classes': irrigation to rich farmers, electricity to industrialists, and good-paying and prestigious work to skilled professionals (e.g., civil servants and engineers)."[87] The less affluent who might suffer because of dam construction were, as mentioned in chapter 1, encouraged to view their suffering as sacrifice in a good cause and to quietly accept it.

The residents who were expected to make sacrifices were not rich farmers, industrialists, and skilled professionals. Rather, they were generally subsistence farmers and members of historically disadvantaged groups. Indian society is stunningly complex. Perhaps the most widely known hierarchical social system in Hindu Indian culture is the caste system, with its multiple strata of vocation-linked divisions. The lowest stratum encompasses the self-named Dalit, once called "untouchables." The Dalits are also referred to by some as Harijan, or "children of God," a term the Dalits themselves find offensive. In the Indian Constitution, the Dalits are identified as Scheduled Castes. There are more than 200 million Dalits in India, more than one-sixth of the population. They continue to work primarily in sanitation-linked occupations, for example, as garbage collectors and road sweepers. Most Dalits live in poverty and are poorly educated. They are often the objects of discrimination and prejudice, sometimes involving violence. Even conversion to religions other than Hinduism does not necessarily provide an escape from Dalit status. Development projects often affect Dalits especially adversely, as is the case with disadvantaged and politically weak groups elsewhere.

In addition to Scheduled Castes, the Indian Constitution also recognizes Scheduled Tribes. This term refers to tribes, known collectively as Adivasi and considered by many to be Indigenous peoples who inhabited the Indian subcontinent before the arrival of the Indo-Aryans and Dravidians. The Adivasi practice a variety of religions, including Animism and Hinduism. The Adivasi population is primarily rural and has been regarded by many other Indians in much the same way as many white Americans view Native Americans and Egyptians regard the Nubians: as primitive and uncivilized. Because their land tenure was unofficial, the Adivasi proved easy targets for resettlement. There are more than 100 million Adivasi, more than one-twelfth of the Indian population. Like the Dalits, they are generally impoverished and poorly educated. They tend to live in heavily forested, often mountainous, areas and practice subsistence agriculture or hunting and gathering. Because they often inhabit areas targeted as prime hydroelectric dam sites, the Adivasi have suffered disproportionate displacement. Their traditional patterns of land tenure (or lack thereof) mean they often receive no compensation for the loss of land. Many end up in India's extensive and growing urban slums.[88]

By the early 1980s, nearly 900 big dams had been built in India.[89] Development came at a cost, though. Ultimately, the process of dam construction in India, which has by no means ended, led to the displacement of more than 40 million people; some estimates are much higher. The numbers are staggering. Moreover, the number of people displaced per hectare today for dam projects is higher than it was in the 1960s and 1970s.[90] Some individuals may have to move more than once, a result of the familiar chaos that often accompanies displacement due to dam construction; they are never really resettled in the sense supposedly intended. At the turn of the millennium the Indian journalist Dilip D'Souza concluded, "Over the years, the people our dams displaced had, without exception, been treated in a manner that brought shame to the ideals and dreams of independent India."[91] S. Parasuraman, a social scientist who has engaged in extensive studies of relocation issues, observed, "The method by which compensation is determined is inadequate, and always disadvantageous to the oustees."[92] He further asserted that what are termed resettlement and rehabilitation policies in India are all "*ad hoc* in nature."[93] *Ad hoc* sounds here like a euphemism for chaotic and uncaring behavior.

Not surprisingly, toward the end of the twentieth century, dam construction in India met with increasing resistance, not only from those directly affected but also from prominent Indian intellectuals like the writer Arundhati Roy and the social scientist and activist Medha Patkar. In her influential essay "The Greater Common Good," Roy denounced big dams in sensational terms: "Big Dams are to a nation's 'development' what nuclear bombs are to its military arsenal. They're both weapons of mass destruction."[94] Domestic opposition to dams in India gained support worldwide, which in some instances helped reduce the enthusiasm of entities like the World Bank for providing financial assistance for big-dam construction in India. The most widely known case of opposition has involved a gigantic project on the Narmada River, which flows through the states of Madhya Pradesh, Maharashtra, and Gujarat. In regard to the Narmada project, the Indian social worker and activist Baba Amte famously asserted: "All those ruthless, monstrous projects Nehru called temples of development have

Displacement and Alienation of Peoples Worldwide   103

become tombs of development."[95] Development sounds here like a form of social death but death that does not have to be passively accepted.

In the 1990s the writer and film director Vishwas Patil (b. 1959) published a novel written in Marathi titled *Jhadajhadati* (1992), which captures in great detail the struggles of those affected by dam construction. An early reviewer explains the title as meaning "a severe compulsory punishment for the sins committed in the previous birth, for crimes that are unknown to the sufferer."[96] When the novel appeared in English in 2014, it was given the title *A Dirge for the Dammed*. Born in the western Indian state of Maharashtra and a former civil servant in the Indian government, in writing the novel Patil drew on his experience as a district resettlement officer in Pune, a city of more than 3 million in Maharashtra. In an interview in 2015, Patil explained how he came to write *A Dirge for the Dammed*:

> [My position] gave me an opportunity to visit several dams. Some of them were in their initial stages of construction while others had been completed. I found that people in those villages were not compensated and were quite indignant about it.
>
> I recollect one particular instance which took place in June . . . About 30–40 people just barged inside my cabin without obtaining prior permission. My memory is still vivid of one old man . . . He looked at me and said that if a lamb is caught in a fire, it is not going to ask anybody in which direction it should jump to save itself.
>
> In that context, he said that their village was surrounded by water from the dam and that they were in a precarious condition.
>
> Such instances with these villagers gradually led to my understanding and empathy with their plight and grief. I felt that they were subjected to heinous conditions and needed a voice to express their views.
>
> Therefore, I decided to be their voice and express these views through my pen. That's how I decided to pen down *A Dirge for the Dammed*. While it is classified as fiction, the story is actually based on real life events.[97]

A Dirge for the Dammed is very much a *roman à these* and an argument against victimization. As such, it explores, in relentless detail—through the experiences of a wide variety of fictional villagers and a fictional river and dam—the many types of trials and tribulations that have been well documented by humanitarian analysts of the plight of peoples displaced by dams worldwide. Dealing primarily with the inhabitants of the village of Jambhli, the novel begins in medias res, ten years after the announcement of the Jambhli Dam Project. For the most part, Jambhlikars are panicked at the thought that implementation of the project may actually begin. The narrator captures their thoughts: "Build the dam and we're all dead. Where should we go now after having lived here for generations?"[98] The unofficial leader of the villagers' agitation against the dam is Vasant Kairmode, known locally as Master or Guruji, a teacher whose slogan is Death before Dam. His background is complex. A member of a Jambhli family, Guruji has spent time away from his native village and seen for himself the impact of the huge historical Koyna Hydroelectric Project on rural people elsewhere:

> The villagers had been forced to move to the upper regions. With no electricity and no money to afford kerosene, they could do nothing but complete their evening meal by firelight. They washed their clothes in stagnant clayey water, the youngsters played catch with sheep and goats and walked eight miles to the new colony to post a letter . . . The dam became operational, Koyna got electricity and Maharashtra prosperity. But the local people on whose land it was built were left in abject poverty. Even after Guruji returned to his village, the anguished, defeated faces of the dam victims haunted him.[99]

In his efforts to avoid a similar post-dam fate for the Jambhlikars, Guruji also pursued an avocation historically peculiar to his family's caste, the Mahar. The Mahar are one of the Scheduled Castes, that is, Dalits. Their traditional role was to protect village borders from nefarious intruders. The narrator says of Guruji: "His family were the official gatekeepers . . . and he had kept up the tradition by awakening his people to the dangers they faced. His love and concern for the Jambhlikars threatened by the dam, his perseverance to get them recompense . . . had made the Jambhlikars forget the nallah [watercourse] that separated him from them. Though a Mahar he had endeared himself to all of Jambhli."[100] Throughout the novel, Guruji practices unyielding, if not necessarily always successful, resistance.

The action of *A Dirge for the Dammed* takes place over many years. The first traumatic event described in detail involves an assault on Jambhli by the police and the forced clearing of a large area that forms part of the dam site. In describing the actions of a large bulldozer, the narrator makes use of an anthropomorphic image that transforms the machine into a monster

Displacement and Alienation of Peoples Worldwide   105

and its natural victim into a maimed and pathetic creature: "The gnarled old babhul [tree] at the edge of the clearing still remained. The bulldozer stood for a moment, licking its lips, taking a good look at its last victim and with all its pent-up power, crashed into its sturdy trunk . . . At the end of an hour a pit had been dug right round the tree. The bulldozer then turned to her roots. The babhul wept sticky white sap, her nerves slashed, her limbs amputated. With a heaving gasp, her mangled body toppled."[101] The villagers themselves are compared to animals about to be slaughtered: "As a sheep in an abattoir sees the butcher's knife reflected in the water it drinks, the people of Jambhli saw spectres of their own future. There was no escape from the brute strength of authority."[102] Patil makes use of similarly evocative and painful comparisons throughout *A Dirge for the Dammed* to convey the social and physical violence wrought upon the villagers.

Much of what the inhabitants of Jambhli encounter involves not only violence but also widespread official hypocrisy and corruption. In Guruji's thoughts, what is touted as the advance of civilization benefits India's elite but not the common people: "The government has devilish powers. As do educated, civilized people. To satisfy their needs and cater to their creature comforts they abuse nature and exploit simple ignorant villagers. But they don't call it exploitation. They give it positive names like progress and development."[103] This is the crux of the problem for Guruji and, in the eyes of many, of mega-dams in general. Although he would prefer to see more modest alternatives to giant dams, Guruji is not, in fact, unilaterally opposed to dams. His concern throughout is for fair compensation and treatment for the villagers, although his spirit of intelligent resistance results in only minimal success.

When the Jambhlikars first visit the town of Khairapur, where many of them are supposed to be resettled, they are initially dismayed by how dry it appears, but they hope the planned irrigation that will follow the dam construction will make the land more productive. The precise site they are initially promised has fertile soil, but the promise is soon rescinded. Both the monetary compensation and the acreage they are offered is less than they had expected. Moreover, their intended neighbors are by no means hospitable: "The residents of Khairapur huddled around doors and windows, grouped in the courtyards and stared, curiosity and contempt mingling in their gaze, at these darveshis [beggars] from some alien land, wandering

106 DISPLACEMENT AND ALIENATION OF PEOPLES WORLDWIDE

gopalas [cowherds], acrobats and fortune tellers . . . One sharp-tongued woman spat at the sight of them, dragged her child inside and shouted . . . 'As if our own troubles were not enough, now we have them descend on us!'"[104] The comparison of the Jambhlikars with landless itinerants exposes the resentment and absence of compassion frequently characteristic of others' view of relocatees, or oustees, as they are often referred to in *A Dirge for the Dammed*.

The actual relocation does not take place for a few years. In the interim, Guruji loses his teaching position because of his opposition to the dam, and his neighbors suffer in multiple ways. "We're all Dalits, slaves of technology, victims of progress," declares Guruji.[105] There are actual deaths. One of the most tragic and symbolic is that of Shivaram, a young man with a new wife and baby who perishes in a bulldozer accident. As in the incident with the tree, the bulldozer here represents the violence of development and technology. When the Jambhlikars are finally able to move to Khairapur, a new series of problems emerge: poor land, extremely inadequate living conditions, uncertain land tenure, family separations, and rapacious and corrupt new neighbors. Occasionally, the Jambhlikars encounter sympathetic and principled officials and regain hope for fair treatment, only to see those hopes repeatedly dashed. Haibati, a young man who for years must engage in insecure labor far from his family, thinks to himself: "We oustees, forget owning bears, we have become bears ourselves, the reins of development strung through our noses. The beneficiaries play the drums of progress and we have to dance."[106] Guruji offers the same comparison as Baba Amte cited above: "They buried us alive and built the mausoleum of the dam over us."[107] The Jambhlikars are even denied their own cremation ground.

Eventually, Guruji's efforts to seek justice excite so much wrath among some of the original inhabitants of Khairapur that a group of young men gangs up on him and beats him up. The assault attracts the temporary attention of journalists, who descend upon Khairapur and publicize the problems of the oustees. Guruji refuses to identify his assailants, concerned that the animosity between the two communities in Khairapur will only worsen. A sympathetic official attempts to call the suspected perpetrators to account but is himself abruptly transferred because of machinations from above. Guruji's daughter is raped by some of the same young men who attacked

Displacement and Alienation of Peoples Worldwide   107

Guruji. Haibati loses his job and returns to his family, only to be forced to demean himself before the original inhabitants of Khairapur to gain even minimal subsistence. His efforts to farm continue to be repeatedly thwarted.

Guruji offers resistance to the authorities until he dies. After his death, another Jambhlikar, young Mhaku, becomes Guruji's unofficial successor and continues to lead the Jambhlikars' struggle for some justice. Haibati is murdered, beaten to death by young men from Khairapur. The narrator laments: "On Haibati's ancestral land, on his razed parental home, the dam stood . . . Such a big dam, such enormous quantities of water, such tremendous progress! With the money that was spent on just one corner of a room of the Jambhli Project Guest House, Haibati's life would have been very comfortable. But progress has fangs. And claws. It has arrogance, but not feelings. No heart, no soul."[108] Progress is a monster with no redeeming features.

At the end of the novel, the Jambhlikars return in force to the dam site—their original home—to protest the crimes committed against them, including the murder of Haibati. The sight of the dam is a huge shock: "What a monstrous structure, crouching on our homes, our lands! They glanced westwards and saw a vast expanse of water. Frolicking tourists enjoying themselves in multi-coloured boats bobbing up and down on it. In these same waters is buried our living, loving, green homeland."[109] The Jambhlikars are convinced to depart, having been reassured once again that proper restitution will take place. The novel ends on an ambiguous note: "There was still much to fight for, to struggle for. A thousand unsolved issues lay before them, like the mountain opposite. Was the battering going to end so easily? So soon?"[110] No, thinks the reader. The same exploitation of the downtrodden by the elites will probably continue, no matter the resistance of the exploited.

A Dirge for the Dammed offers a lengthy and painful narrative. In addition to providing a virtual and vividly detailed catalog of the problems associated with the displacement of people in the service of large dam construction, Patil also eloquently communicates the fact that relocation issues are not a brief and temporary phenomenon. Rather, the impact of displacement goes on for decades, perhaps forever. In Patil's handling, the myth of technological progress is repeatedly exposed as a gain for the well-off and an ongoing loss for the impoverished.

The Chinese and the Yangtze

The Chinese have fought to control rivers for millennia. Horrific floods have long periodically devastated vast riverine areas in China and caused tremendous loss of life.[111] The Yangtze, a large and long river that flows from west to east across central China, provides a compelling example of the costs of flooding. In 1931, 150,000 people perished because of flooding, 8.5 million acres were flooded, and 10 million people lost their homes. In 1954, more than 33,000 people died and almost as much acreage was flooded; in 1999, 13 million people had to be evacuated, and at least 3,300 people died.[112] A desire to prevent such disasters has been accompanied by a longing to improve navigation on treacherous portions of the Yangtze and a strong interest in the potential benefits of hydroelectricity.

Perhaps the best-known and most controversial Chinese large hydroelectric dam project is the Three Gorges Dam, which is located on the Yangtze in central China near the town of Sandouping. The three gorges in question are the Qutang, Wu, and Xiling Gorges. The installation of the dam, which was discussed—often heatedly—for decades, was finally completed in 2012; it is the largest hydroelectric power station in the world. Sun Yat-Sen, the first leader of the Nationalist Party of China, wrote in the early twentieth century about the possibility of a large dam near the Three Gorges. Specific plans for construction began to be developed in the 1930s but were interrupted first by the Second Sino-Japanese War, then by World War II, and finally by the Chinese Civil War. The United States Bureau of Reclamation engineer John L. Savage, who played a major role in the construction of several major American dams, was involved in planning in the 1940s; after the Communist Revolution of 1949, for many years during the Cold War, the Chinese turned instead to the Soviets for assistance with large dam construction.[113]

One of the more sensational moments in the saga of Three Gorges took place after the 1954 flood, when Mao Zedong swam across the Yangtze and subsequently produced a poem outlining his vision for the river: "Great Plans are being made; / A Bridge will fly to join the north and south, / A deep chasm will become a thoroughfare; / Walls of stone will stand upstream to the west / Till a smooth lake rises in the narrow gorges."[114] Mao's enthusiasm was by no means completely altruistic. As the British American journalist Simon Winchester comments, "The dam was seen by Mao and his allies as perfect propaganda for the promotion of his authority and power."[115] This was very

Displacement and Alienation of Peoples Worldwide 109

much in keeping with the tradition elsewhere in the world—Egypt and India, for example—of dam propaganda as an elevation of national prestige.

Actual construction of the Three Gorges Dam was long delayed, however, in part because of the building of another dam, the Gezhouba, downstream from where Three Gorges was eventually located but also because of serious concerns and disputes about the dam's viability. Construction finally began in the 1990s, with extensive Western involvement and financial support but also opposition. Domestic opposition to the dam was silenced. The journalist Dai Qing, for example, who had published critical observations by Chinese engineers in the collection *Yangtze, Yangtze* (1989), was arrested and imprisoned for almost a year in the aftermath of the Tiananmen Square furor. She continued to collect criticisms of the dam, many anonymous; they appeared in the collection *The River Dragon Has Come!* (1998). The pieces bear witness to an urge for large-scale development willing to ride roughshod over human interests. Some of the most painful observations are related to the collapse during a horrendous typhoon on August 8, 1975—due in part to shoddy and hasty construction—of the Banqiao Dam on the Ru River, the Shimantan Dam on the Hong River, and sixty other dams in central China; as many as 240,000 people died. As for Three Gorges, the historian Judith Shapiro observes elsewhere that the dam is considered by some "to be a monument to Communist Party hubris."[116]

The concerns about the Three Gorges Dam were and are typical of concerns about large dams, but they are often greatly magnified because of the dam's immense size. In *Green China: Seeking Ecological Alternatives* (2002), Geoffrey Murray and Ian G. Cook provide a detailed and specific list of probable environmental problems resulting from construction of the dam.[117] The problems include siltation and sedimentation, pollution, deforestation and erosion, endangerment of wildlife, and earthquakes triggered by the weight of the dam. There were also worries about the impact of possible bureaucratic corruption and vulnerability to terrorism. The loss of multiple archaeological sites was lamented. Then there was the relocation of people. The numbers are staggering; an exact count is unavailable, but at least 1.25 million residents had to move, in some cases far from their original homes. What happened with relocation in India has been more than matched in China.

Given the authoritarianism of the Chinese regime and its commitment to the Three Gorges Dam, it will come as no surprise that novels critical of the

110 DISPLACEMENT AND ALIENATION OF PEOPLES WORLDWIDE

dam have thus far only been published outside China. A prime example is Li Miao Lovett's *In the Lap of the Gods*, which was first published in English in 2010. As a child, Lovett immigrated to San Francisco, where she has been actively involved in environmental and social issues.

In the Lap of the Gods offers a devastating portrait of displacement and resettlement issues linked to Three Gorges (figure 3.2). The novel's major protagonist, Liu Renfu, is an illiterate day laborer who at times, to make a minimal living, has to resort to scavenging for potentially valuable items in abandoned riverine areas that will soon be flooded by reservoir waters. Throughout the narrative Liu continues to suffer from grief over the drowning deaths of his first wife and their unborn child in a ferry accident. Marriage to his second wife does not bring happiness to either partner and ends in separation. What keeps Liu going, through poverty, physical accidents, and eviction, is the presence in his life of a child he finds as an infant at the beginning of the novel and eventually names Rose. Liu discovers Rose while scavenging in what evokes a post-apocalyptic atmosphere: "The waves lapped against the new shore, muffling the baby's cries. As the water advanced, it threatened to swallow the wicker basket resting on a spur of limestone. The river, now a growing lake, crept up the fields inch by inch . . . The baby in the basket squeezed her eyelids shut and cried . . . The last inhabitant had cleared out several hours earlier."[118] Liu observes the departure of the baby's parents; when he comes upon the crying infant, with instinctive sympathy he rescues her from imminent drowning. Initially, Liu considers turning the baby over to an orphanage for financial compensation. Later in the novel, by a series of coincidences, he is contacted by the baby's grandmother with an offer of money for the little girl. Liu does not relinquish Rose, however; "she was the nymph spirit that had kept him from giving up."[119]

Another major character in *In the Lap of the Gods* is Fang Shuping, an old man who makes his living as an unofficial broker of deals and is Liu's fence for his scavenged items. Although Fang initially appears to be grasping and selfish, in the course of the novel a more positive side of the old man is revealed. He has in the past attempted to assist peasants seeking to claim their rightful share of resettlement funds due to dam construction, and later in the novel he does so again, largely because of a possible reconnection with a woman he loved and was forced to separate from decades earlier. Fang

Displacement and Alienation of Peoples Worldwide 111

FIGURE 3.2. Christoph Filmkössl, *Three Gorges Dam*, photograph, 2006. Wikimedia Commons, https://commons.wikimedia.org/wiki/File:Dreischluchtendamm_hauptwall_2006.jpg.

perishes, though, when he is caught up in a violent skirmish with soldiers during a resettlement protest by peasants.

In writing *In the Lap of the Gods*, Lovett's research was informed by onsite visits, expert consultants, and publications from the nonprofit environmentalist organization International Rivers. The novel is filled with incidents and stories relating to demonstrations and unsuccessful attempts by peasants to gain compensation, corruption of officials, dismantlement of villages to be flooded, and impoverishment and family separations in the wake of displacement. Liu is present when the old town of Wushan is dynamited in preparation for flooding. Residents are racing to leave, except for one old woman who moves along slowly. When Liu asks her where they are relocating, she replied: "'Far away.' She blinked and then stared at him with her tired, sunken eyes. 'I would rather die and be buried here, but this will be a town of ghosts.'"[120] Like the elderly characters in other works discussed in this chapter, the changes wrought by the dam construction overwhelm this old woman.

112   DISPLACEMENT AND ALIENATION OF PEOPLES WORLDWIDE

When the demolition takes place, the scene evokes a war zone: "Moments later, a great explosion rocked the hillside. The old city trembled, and then it began to tumble to the earth—tall buildings tottered and fell like drunken men, shells of low-rises crumbled and turned to dust . . . Smoke rose from the city in torpid black clouds as explosion after explosion went off. Old Wushan had become a ship tossed at sea, and when the storm subsided, the city would sink into a remorseless, watery grave."[121] There are several descriptions in the novel of protests by relocatees trying to acquire fair financial compensation for their removal and related disruptions to their existence. All of these incidents end in police violence. Liu is by chance present at one such event:

> "Break it up!" A dozen nightsticks swooped through the air in an acrobatic arc before descending on the heads of the *lao bai shing* [ordinary people]. All around, heads swayed and bodies crumpled to the ground. The jumble of arms and legs flailed against an indifferent sky.
>
> The nightsticks flew in ten directions. Protesting heads fell silent. Uncombed heads fell into disarray, surrounded by the gleam of silver-capped teeth, and the grizzle of unshaved beards. Frightened heads cried for mercy, but the nightsticks soared into the air and swooped down like buzzards.
>
> A bristly head near Liu sank to the ground, and cracked teeth scattered on cement, bathed in a crimson pool. A wispy gray head, fragile as a dandelion puff, surrendered to the lamppost, as ancient legs fell against Liu's heels.[122]

Such scenes capture the poverty, frequent old age, and extreme vulnerability of the millions displaced by the construction of the Three Gorges Dam. *In the Lap of the Gods* offers a searing condemnation of supposed progress but also reveals the relocatees' refusal to accept change without a fight. Liu Renfu's adoption of Rose is symbolic of a continued spirit of a love of life.

There are many thematic commonalities in the novels and stories discussed in this chapter. The poignant loss of traditional domestic beauty captured in *Farewell to Matyora* is evoked, for example, in *Dirge for the Dammed*. The violence that marks the dismantling of the cemetery and uprooting of trees in *Farewell to Matyora* is echoed by the bulldozers in *Dirge for the Dammed* and the dynamite in *In the Lap of the Gods*. The particular anguish experienced by the elderly is captured in Rasputin's Dar'ia, Ali's Hushia, and many of Patil's and Lovett's characters. The shoddiness of the new settlements

Displacement and Alienation of Peoples Worldwide 113

of the displaced described in *Farewell to Matyora* and "The Fire" is matched by the sordid living conditions of the villagers in "Adila, Grandmother" and "The River People" and the Jambhlikars in *Dirge for the Dammed*. The post-displacement erosion of financial security experienced by several characters in *Aurelia* resounds loudly in *Dirge for the Dammed* and *In the Lap of the Gods*. The breakdown of social ties and sense of community that marks "The Fire" assumes horrific proportions in *Dongola* and *Poor*. The infliction of violence on the vulnerable displaced that occurs in *Dirge for the Dammed* continues in *In the Lap of the Gods*. The corruption and untrustworthiness of the authorities plays a prominent role in *Aurelia*, *Dirge for the Dammed*, and *In the Lap of the Gods*. All of the works treated bear witness to the analyses by social scientists and assertions of journalists and humanitarian environmentalists. Displacement causes poverty, poorer living conditions, the breakup of families, corruption, diseases, a need to engage in migrant labor, and a decline in the position of women. Literary works bring such problems alive in a meaningful, poignant, and convincing manner.

It is also apparent from a consideration of the works discussed here that there has been tremendous worldwide growth in the problems of human displacement and the numbers of people displaced because of the construction of big dams, from the United States to Africa, India, China, and Russia. One could list many other countries similarly affected. What might once have seemed lamentable but localized has increased in scale to the point that it appears to constitute an existential constant. Many of the characters in the novels and stories analyzed retain a sense of dignity and hope for the future and at times act on their frustration, but the narratives themselves afford little cause for optimism. And so the process of dam building and displacement and alienation continues, touching and irreparably damaging the lives of millions. Literature can and undoubtedly will, as it already has, expose this situation in dramatic and humanizing ways that purely scientific treatments cannot do.

4

Contaminated Water, Disappearing
Fish, and Deadly Sediment

-By the end of the twentieth century, the detrimental environmental consequences of large dams had emerged more clearly and in more diverse ways. From the aesthetic to the transformational, from disease to ecocide, a host of effects came to light—some anticipated but underestimated or ignored, others unforeseen but nonetheless important. By no means was only fishing affected. Many of the literary works treated in chapter 3, as well as other writings not yet discussed, explore these issues. Valentin Rasputin (1994) brings to life the unpleasant experience of brackish water on the Angara, Bill Gulick (2008) traces the uphill battle to maintain the salmon population of the Columbia River, Elizabeth Cook-Lynn (1999) describes the barren shores of the dammed Missouri, Idris Ali (1998, 2007) writes of the Nile mud laden with bilharzia parasites, and Alex Kuo (2011) looks at sedimentation on the Yangtze. While these topics may not at first glance seem likely to excite dramatic fictional engagement or probing literary analysis, writers worldwide have captured their trauma and significance in ways that engage readers' attention and deserve concern. The impact on humans of environmental

https://doi.org/10.5876/9781646425976.c004

problems related to dams looms large in these works and often overlaps with the problems of displacement and alienation discussed in chapter 3. The literary allusions to and treatment of specific environmental issues, however, very much warrant a closer look in and of themselves. An analytical focus on literary treatments of dam-linked environmental areas highlights the variety and cascading impacts over time of the damage to nature that may be caused by large dams. The author and scientist Mark Everard argues that "large dams do not so much modify ecosystems as represent a wholesale replacement of one set of ecological services with another."[1] Fictional writings support the painful validity of this assertion of complete ecological transformation and its impact on nature and human beings in an often moving or dramatic manner.

A dam reservoir has an appearance very different from that of a natural lake. The land edge begins abruptly, lacking the transitional quality of a riparian zone. Because of the periodic engineered lowering and raising of a reservoir, the barren land that borders it may be subject to erosion. In its early years, rotting trees and other vegetation may fill a reservoir; the desirable removal of ground cover before the filling of a reservoir is sometimes perceived as too time-consuming or too costly, and builders are often in a rush to reap the financial benefits the construction of big dams affords. The bedraggled appearance of many reservoirs does not typically evoke exclamations about their beauty. But such visages are only the beginning, the tip of the iceberg, an aesthetic failing that conceals highly detrimental environmental consequences.

In the past few decades, many environmentalists, journalists, and academics have explored in detail the environmental ramifications of large dams. Two of the most thorough treatments are found in *The Social and Environmental Effects of Large Dams* (1984) by Edward Goldsmith and Nicholas Hildyard and *Silenced Rivers: The Ecology and Politics of Large Dams* (2001) by Patrick McCully.[2] Goldsmith and Hildyard identify a host of problems, beginning with perhaps the most obvious: the loss of prime agricultural land. So-called bottomland is often famously fertile; one thinks of Sergeant York's moving desire, highlighted in the 1941 film of the same name, to purchase a piece of bottomland. Productive forests may also be lost to reservoirs and lumbering, along with the wildlife that inhabit them. Downstream from dams, the beneficial impact of nutrient-rich silt disappears; in a break with millennia-long tradition, Egyptian agriculture ironically now requires increasing amounts of artificial fertilizers. The introduction of perennial irrigation that often accompanies the construction of a dam may attract organisms and insects inimical to both plant and human life. Malaria, schistosomiasis (bilharzia, or snail fever), and the less commonly known filiariasis (elephantiasis) and onchoceriasis (river blindness) may flourish, particularly in tropical and subtropical regions. Fish populations change and diminish. Riverine fish often do not tolerate stagnant reservoir conditions well, and the decrease of water level downstream—often accompanied by increased salinization—can also have an adverse effect on fish populations. Herbicides and pesticides may be used in greater quantities in reservoirs, contributing to a decline in water quality and biological damage to fish. Dams may collapse, and the pressure of reservoir water may give rise to earthquakes. Finally, there is sedimentation, which can literally kill a river: "Sooner or later, the reservoir of a dam must fill up with the silt and other detritus that the dam prevents from flowing downstream. And when that happens, the dam must be decommissioned; without its reservoir, a dam is a useless slab of concrete."[3] This will not happen in the distant future; a life span of 100–200 years for a large dam would be unusual, 50 more is typical, and there are many examples of dams that have demonstrated a much shorter viable existence.

McCully touches on many of the same issues Goldsmith and Hildyard cover. He emphasizes the dramatic ways upstream dams may damage river deltas and drastically lower the amount of soil moving downstream. Of the Colorado River delta McCully writes, "The indigenous Cucapa, or 'people of the river,' who once fished, farmed and hunted the delta, have been reduced from a population of 1,200 a century ago to just 40 or 50 families barely subsisting on a diet of beans and junk food."[4] He emphasizes that the ecology and hydrology of rivers still often continue to be poorly understood before they are dammed and that it is risky to extrapolate in the tropics on the basis of experience in temperate regions; a major problem, for example, is a higher rate of evaporation from reservoirs in warmer climates. McCully also points to the unusual diversity of the river and floodplain habitats reservoirs destroy and the deleterious impact of stilling flowing water. This may not be immediately apparent, but there have been egregious examples of noxious gases emanating from young reservoirs, particularly in Latin America. As for anadromous fish like salmon and steelhead trout, the list of dangers with which dams threaten them has become longer and longer and the number of fish living in dammed areas smaller and smaller. This has had a pronounced economic effect; for example, "the National Marine Fisheries Service have [*sic*] estimated the cost of salmon fishery losses due to dams in the Columbia Basin for the period between 1960 and 1980 alone at $6.5 billion."[5] Native Americans have accrued much of that loss. McCully concludes his overview of the environmental effects of dams by denouncing the argument sometimes heard that environmentalism is nothing but a luxury for the affluent: "An argument often used by dam builders and backers in developing countries to defend incomplete and biased environmental surveys is that concern for the environment is a 'first world luxury' . . . In fact the opposite is the case. The majority of people in developing countries depend directly on their environment to provide them with subsistence. The environmental destruction caused by dams . . . thus carries a major human cost, which falls most heavily on the poorest sections of society."[6] Literary works support this contention in an aggressively dramatic and effective manner.

In recent years, environmental analyses have paid greater attention to dam-related carbon emissions and their role in global warming. In *Concrete Revolution*, Christopher Sneddon points out that research into reservoir emissions did not begin until the early 1990s and even now is often subordinated to the unexamined assumption that hydroelectric dams serve as a clean source of energy far superior to fossil fuels.[7] This is just the latest example of a touted benefit that conceals a long-term danger. As will be seen, such slowly emerging risks are receiving increased attention in world fiction, especially in the case of recently constructed and controversial large dams like China's Three Gorges Dam on the Yangtze River.

Lamenting Dirty Water in Russian Fiction and Poetry

The hoopla associated with the construction of dams on the Siberian Angara River, especially the gigantic Bratsk Dam, was discussed in chapter 2, and the issue of relocation of Russian villagers on the river was covered in chapter 3. Valentin Rasputin, who has compellingly evoked the social effects on Siberian villagers of relocation because of dams in his writings, has also highlighted environmental consequences in both subtle and graphic ways, especially in his essayistic story *"Vniz i verkh po techeniiu"* (Downstream and Upstream, 1972).[8] What happened with the Angara reflects many of the environmental problems cited above. For example, hundreds of thousands of acres of forest along the Angara were supposed to provide useful timber; ultimately, because of insufficient time and not enough roads, much of the forest area was simply flooded. This resulted in problematic navigation and an excess of decaying matter in the water, which, in turn, had a deleterious effect on fish. The many factories that took advantage of the electricity generated by the Angara dams also proved to be a source of serious pollution. Western scholars eventually declared the Angara "an aqueduct for poisons."[9]

In "Downstream and Upstream," the first-person narrator, the writer Viktor, begins his journey to his family's home on the Angara with nostalgic daydreams about the beauty of the island landscapes of his youth. Suddenly, though, Viktor remembers that these islands no longer exist. A vague sense of anxiety replaces pleasant memories. The next morning Viktor wakes up to a strange and disturbing sound, caused, it turns out, by branches of trees standing in the water scraping against the ship. Going on deck for a better look, Viktor is confronted by a disturbing sight:

> The ship was fighting its way inside some sort of mysterious, wide bay, from whose shores trees extended far into the water. They were sticking out both ahead and behind. It was unclear what was more surprising and improbable: whether to consider the ship, carefully crawling among the trees, a giant prehistoric monster or to look at the trees growing out of the water as if at a fantastic painting.
>
> But was it fantastic? The trees were naked and pitiful, leafless, with sparse, furled needles, with slimy branches swollen from the water, with black catkins resembling caterpillars on the birches. Some trees were still standing straight, others were already bent over, they were being slowly washed away. The waves, bubbling up and down, rode up on them, and then the trees shook with a sniveling moan.[10]

Eventually, this dreary and repellent sight, to which literary narrative gives particular intensity, yields to an open expanse of gray and motionless water: the reservoir, which the locals now refer to as a sea. This is an important linguistic difference; the river is no more and the "sea" cannot offer the same

Contaminated Water, Disappearing Fish, and Deadly Sediment 119

riches as a river. The reservoir water indeed seems curiously lifeless, lacking the play of colors, movement, and sounds of the former river. This is truly a pathological landscape, one that emanates death and a feeling of despair.

At the new location of his former village, Viktor discovers that the aesthetic shock the reservoir affords is only the first of multiple disappointments. While there are many fish in the reservoir, they are inferior in quality compared to the edible fish previously available, as a conversation between Viktor and one of his relatives reveals: "There are a lot. But what kind of fish: perch, roach, pike. The pike are as healthy as logs. But their flesh isn't quite right, it's actually kind of wooden, and tastes of slime. We even feed it to the pigs." "What about grayling and trout?" "From where? Those are clean fish and they need clear, flowing water. For grayling you need to go up to the headwaters of the upper reaches of the Kui."[11] In addition, the reservoir water is not potable or otherwise especially usable. When Viktor's mother sets out with buckets for water, he expresses surprise because there is a full cistern of water right at the hut. His mother explains: "This water isn't good for tea. It's from the water tower, and it's very hard. You can't wash your hair with it, for example, you can't get any kind of comb through it then. We still use it for slop, but for tea we lug water from the river, even though it's far away."[12] This is much more than simple inconvenience. The water has greatly declined in quality, causing a host of serious problems that gradually emerge in the course of the narrative.

Viktor offers to get the water himself. Down by the reservoir, he realizes that acquiring suitable water for tea, however, is more complicated than he had realized. The water near the shore is red and clayey and looks undrinkable. Viktor walks a few feet out into the water and discovers that the ground below is disgustingly slimy. To Viktor's surprise, a boy informs him that to obtain freshwater, it is necessary to go out even further from shore in a boat. He offers to assist Viktor. After they get the water, the boy suggests a swim. Now Viktor learns how troublesome and even dangerous the rotting trees under the water can be. The boy warns him not to dive in a certain spot: "There's a pine tree somewhere under the water there. Last year Mishka Zhukov dove there and ripped up his whole side. There was blood. Somehow we got him to the shore."[13] The tree wounded Mishka so badly that he had to have an operation. Through such incidents, Rasputin effectively manages, without engaging in an explicit diatribe, to convey the manifold variety of

120 CONTAMINATED WATER, DISAPPEARING FISH, AND DEADLY SEDIMENT

visual and physical environmental damage the dam has already wrought to the water, the living creatures who inhabit it, and the humans who live beside it. This is a tremendously effective use of literary description.

During the period of glasnost and perestroika in the Soviet Union and in the aftermath of the country's breakup, writers for a time were more readily able to express criticism of environmental damage, a previously forbidden topic. The Volga, with its lengthy cascade of big dams, was a particular object of discussion. In the memoiristic piece *"Plyvushchii korabl'"* (Sailing Ship, 1989), the literary critic Marietta Chudakova reminisces about a trip she took on the Volga in 1967 and describes her appalled impressions at the sight of a giant reservoir:

> The shores disappeared. We were no longer sailing along a river, but along a strange, boundless, smooth surface, neither maritime nor riverine. Of course I knew about the dam and the reservoir, but for some reason I hadn't expected such a harsh impression. I well remember that the feeling that arose was not exhausted by bitterness, but to a much greater extent it was a powerless anger. We were sailing along a river that had been taken away from a great people, and it was impossible to rid oneself of the excruciating sense that no one would ever be able to look at what was drifting by, at the invisible shores, with the eyes of those who gazed at them 60, 100, 200, and 300 years ago. I couldn't manage to rid myself of this feeling—too much had been taken away, and somehow shamelessly, in an impersonal way.[14]

Chudakova's statement echoes the concerns of Rasputin's narrator. This reservoir is not a river or even a lake but a monstrously unnatural human creation.

Many other Russian intellectuals and writers have echoed Chudakova's grim impressions. In 1990 a collection of essays titled *Stony Volgi* (Moans of the Volga) appeared. Several of the contributors lamented the condition of the river. Viktor Drobotov's anthropomorphized mother, Volga, begs for his intervention: "If you have a conscience and sense of civic responsibility, you must not, you do not have the right to be silent about what you have seen with your own eyes."[15] For Aleksandr Tsukanov, the well-being of rivers is also linked to that of humanity: "The Volga is not a toilet, not a bottomless barrel, and not a streetwalker, but one's mother . . . on her health depend the health and life of an entire nation."[16] Here again is the familiar Russian metaphor of river—particularly the Volga—as mother, as something to be cherished and loved, not tamed and dominated.

Another volume of essays on the dire Russian ecological situation, *Ekologicheskaia al'ternativa* (The Ecological Alternative), also appeared in 1990. Like the American writer Frank Waters, who came to regret his earlier enthusiasm for the Hoover Dam, one of the contributors to *The Ecological Alternative*, the poet and author Nikolai Pal'kin, has had a serious change of heart about the Volga's constrained and polluted fate. Pal'kin explores the metaphor of river as mother even more profoundly and depressingly than did Drobotov and Tsukanov: "And if I write about the Volga again, then it's not like yesterday. Not only with delight, but also with spiritual pain. I catch myself with the thought, or more precisely, the feeling that to speak about the great river as we spoke not long ago at all is to be cunning, or worse than that, to speak a half-truth. It's the same as talking about the health of one's mother while concealing her serious illness."[17] His poetic conclusions are even darker: "*Pozhalui, i zhdat'-to nedolgo / Kogda uspokoitsia mat'. / I budem v pechali nad Volgoi / My vse, kak nad grobom, stoiat'*" (Perhaps we don't have long to wait until our mother is laid to rest. And we will all stand in sadness above the Volga as if over a grave).[18] Like Rasputin, Pal'kin implies that there has been a profound moral failure where Russian rivers are concerned.

Trying to Save Salmon on the Columbia River

As is clear from the discussion of Craig Lesley's novels in chapter 3, *Winterkill* and *River Song*—especially the latter—focus in large part on the cultural implications of Native American losses of traditional fishing sites on the Columbia River. The fact that not only have such sites vanished but that the number of fish on the river has precipitously declined is occasionally mentioned or implied in Lesley's narratives. In a conversation early in *River Song* between the protagonist (the Nez Perce Danny Kachiah) and a pear orchard owner, the owner alludes to the days when the salmon canneries on the river were "in full swing."[19] At the Wind River fish camp, one of the in-lieu sites the government provided to the Indians of the Columbia River Basin after the construction of the large dams, the old former Celilo resident Willis Salwish tells Danny: "We've been getting some pretty good silvers [salmon] . . . Kind of slow, though. Summer runs were big before the dams."[20] How much bigger Willis's sharp-spoken granddaughter Velrae makes explicit in response to a comment by her grandfather about the Creator sending the salmon: "Thanks to the dams, He only sends about a tenth of the salmon He used to."[21] Such passing but nonetheless striking (and accurate) comments point to the immense toll taken on the salmon population of the Columbia in the second half of the twentieth century. This has become a huge economic problem, and not just for Native Americans.

The fish losses mentioned in passing in *River Song* are explored in exacting detail in Bill Gulick's documentary novel *The Dam Builders* (2008). Grover C. "Bill" Gulick (1926–2013) is the author of the trilogy *Roll on, Columbia* (1998), whose title is taken from Woody Guthrie's famous song of the same name mentioned in chapter 2. Gulick's work is a massive historical saga that treats the development of the Columbia River Basin from the efforts of the early nineteenth-century settler colonialists to those of the mid-twentieth-century dam builders. Meticulously researched and exhaustive in its detail, the trilogy re-creates historical events through the activities and actions of the fictional Warren family. The third novel in the trilogy, *Into the Desert* (1998), ends with the Vanport Flood of 1948 near Portland, Oregon, whose financial and human costs provided partial impetus for more extensive development of a system of dams on the Columbia encompassing flood control, improvement of navigation, and the generation of hydroelectric power. In the epilogue to *Into the Desert*, Gulick observes that "one of the major unanticipated results of the dam-building era was the damage done to the fish runs" and briefly outlines the complex components of the problem.[22]

The Dam Builders picks up the story of the Columbia where *Into the Desert* leaves off and pursues it into the late twentieth century. A major topic of the novel is the fish losses on the Columbia. The main protagonists of *The Dam Builders* are Larry Scott, a World War II veteran who loses his first wife and son in the Vanport Flood, and Gayle Warren, daughter of a geologist and a biologist. The two marry in the late 1940s. Larry works as an electrical engineer for the Bonneville Power Administration, Gayle as an ichthyologist (fish specialist) for the Army Corps of Engineers. Concerns about Native American fishing rights figure repeatedly in the novel, but great attention is also lavished on the details of the closely related fate of the fish on the Columbia. Early in *The Dam Builders*, Gayle's father says that his daughter and his wife, Jenny, "think the dams the government is planning to build are

Contaminated Water, Disappearing Fish, and Deadly Sediment   123

going to destroy the runs—and I suspect they're right. Gayle and Jenny are determined to do all they can to save the fish."[23] This is a much more complex undertaking than is initially apparent.

Gulick's novel was written with the benefit of decades of hindsight, so that, for example, the fish ladders included in some Columbia River dams—such as Bonneville Dam—of which wondrous things were expected in the 1940s, are treated with a certain skepticism from the beginning of *The Dam Builders*. Gayle studies the impact on the fish of the ladders and turbines at McNary and the Dalles Dams. A significant percentage of fish, it turns out, do not make it up the ladders that were supposed to be so helpful. The next solution pursued was upstream fish hatcheries, which would, many experts thought, compensate for the loss of fish heading upstream through the ladders and for the fact that some dams, such as Grand Coulee, were too high for fish ladders; hence, there was no way for the fish to continue upstream beyond them. But hatcheries brought their own problems.[24]

Other difficulties that soon emerge with fish on the Columbia are discussed at length in *The Dam Builders*. Fish heading downstream die in large numbers. The culprit, it is discovered, is gas bubble disease, the "bends," caused by nitrogen that can accumulate in the pools beneath dams. Gayle and her boss, the historical head of the Fish Preservation Department of the Corps of Engineers Ivan Donaldson, discuss the problem and possible solutions. These include transporting smolts by tanker trucks or hauling them by tug and barge. Barring sports fishing for a few seasons is also used as a means of providing the fish with a temporary respite. Gulick's novel gives much attention to the lack of enthusiasm and even anger such measures engender in some quarters, where the central concern is access to recreational fishing. Another unanticipated problem that is later revealed is a particular vulnerability that hatchery fish exhibit: "Hatchery-raised fish, the biologists learned, do not possess the natural immunity to disease that wild fish do."[25] Gayle spends her entire career learning about, explaining, and trying to confront these complex problems.

After retirement, Gayle plans to work with tribal fish biologists on trying to restore lost salmon runs. In some instances, such attempts succeed, but it is an uphill battle. Near the end of *The Dam Builders*, Gayle tells her husband about the conclusions of an actual extensive report on the essential problems for fish on the Columbia: "What they amount to . . . are *Harvest, Habitat,*

124 CONTAMINATED WATER, DISAPPEARING FISH, AND DEADLY SEDIMENT

Hydro, and *Hatcheries*. Unless we solve these four problems . . . all eight species of anadromous fish runs in the Columbia River system are doomed to extinction within fifty years."[26] Reading *The Dam Builders* is a rich educational experience. Gulick succeeds admirably in presenting a serious environmental problem, one with financial implications, in a thorough yet accessible manner.

In *Winterkill* and *River Song*, Craig Lesley shows the pain of Native American losses of traditional fishing sites on the Columbia River. In the character of Danny Kachiah, Lesley hints at the possibility of a renewed and meaningful commitment to traditional ways. In *The Dam Builders*, Bill Gulick outlines a careful scientific path to the possible restoration, at least to some extent, of the decline of salmon on the Columbia. In *K Falls* (2001), Skye Kathleen Moody paints a very different, and darker, picture of the salmon dilemma. This novel will be discussed in chapter 5.

The Missouri Defaced

Fish are rarely mentioned in *Aurelia: A Crow Creek Trilogy* (1999), Elizabeth Cook-Lynn's chronicle of the impact of Missouri dams on the lives of Sioux Indians from the 1950s to the 1990s. Environmental consequences are sometimes portrayed in very general terms, using formulations like "the ruined river" and "the destruction of the river," and at other times through verbal pictures of devastation—reminiscent of Rasputin's descriptions of the dammed Angara—that evoke the ecological harm that has been done.[27] A sense of death, sterility, and predation marks such descriptions: "The river lay in the distance, darkly lapping at the barren shore. Withered, whitened cottonwoods sank into the cold sandy shorelines and coyotes were the only silent prowlers on the land"; "they would haul a few dead trees lying along the shore of the ravaged river into their yard"; "treeless, the river seemed like a gigantic misplaced waterscape on a level with the brown prairie grasses now turning brittle"; "at last Aurelia felt all right about why she wanted to weep every time she saw the dead trees still and white, the dams and the swift water and the seething foam made at the spillways. The vast emptiness caused by the destruction of the river."[28] References to destroyed timberlands and vanished nutrient-rich river bottoms allude to one of the most obvious and immediate, but nonetheless crucial, consequences of damming. The overall impression is one of destruction and death.

Contaminated Water, Disappearing Fish, and Deadly Sediment   125

As the trilogy progresses, the characters' thoughts about the river increasingly reveal a specific and knowledgeable environmental understanding. Aurelia, the major protagonist, muses as she looks at the Missouri: "The dammed river was in the distance and its surface shone like crushed diamonds, but Aurelia knew its gleam to be hiding a terrible ruin . . . Used to be a great river, she'd say over and over. She wondered about what was told by some, that the muddier the river, the sooner a dam will silt up, topographies erode, and fish stocks be damaged. What, she wondered, did all that mean to the future?"[29] Aurelia is well informed, but will that matter? Will she and other Native Americans be able to apply their knowledge?

An emphasis on environmental damage is especially pronounced in the third novel of the trilogy, *In the Presence of River Gods*, which is set in the early 1980s, with an epilogue in 1995. The narrator's preface to the novel casts riverine pollution in broad terms marked by ethical implications: "The Hindus of India and Bangladesh still immerse themselves in the Ganges River, in spite of the fact that it is filled with toxins, human remains, and garbage . . . The Dakotapi of North Plains still believe in the *Mni Sosa* [Missouri] spirit who lives with them in their river country, even though the [Missouri] and the land [bordering the river] have been destroyed by hydropower dam development and technology. They know that to lose contact with the river gods is to lose everything."[30] The comparison between the Ganges and the Missouri points to the spiritual quality a river may possess for human beings and the way development can undermine that spirituality.

As discussed in chapter 3, the central event in *In the Presence of River Gods* is the rape and murder of a young Indian girl, an event the narrator and the novel's Indian characters link to both racism and the destruction of the Missouri. Symbolically, the girl's body is found in the devastated area that borders a reservoir. The description of destruction and pollution that appears at this point in the trilogy is by far the most lengthy and traumatic:

Fluctuating reservoir waters had been rising and falling for more than two decades along this river and they often left slick muddy beaches or soft, devastating pollution in alluvial soils. For eight hundred miles . . . aquatic vegetation and huge timber stands had been destroyed by the monster dams and reservoirs. Bacteria attacked whatever the water held for any period of time and quick destruction of the natural world was noticeable . . . North

126   CONTAMINATED WATER, DISAPPEARING FISH, AND DEADLY SEDIMENT

of town and along the shoreline where the body was found, the waters had grown thick with fallen cottonwoods, and farm and ranch dogs lapping along the shore often limped home only to drop dead with swift and unaccountable infections . . . Fallen cottonwoods, all manner of driftwood and tree roots littered the shore for hundreds of miles, looking for all the world like the war carnage left after a massive battle. Stiff skeletons of trees still stood as surviving clumps up and down the river even twenty years after their roots were drowned.[31]

This horrific description captures both the visible and invisible ruin of the Missouri, which is now a kind of vicious hell. This is by no means the narrator's last word. The river does ultimately have its revenge: "In the 1940s, just after the Good War, the offspring of these early immigrants, along with their federal government, took control of an encouraging world by damming up the Missouri River to keep it at bay. Half a century later, they could do nothing about the dogged Red River's intention to get its revenge for all the rivers in the world when it devoured the carefully nurtured towns in the worst flood ever recorded."[32] Cook-Lynn is alluding here to what is known as the Great Flood of 1993, which occurred throughout several states along the Missouri, the Mississippi, and their tributaries and lasted several months, causing $15 billion in damages. There were multiple immediate causes of the flood, most notably persistently heavy rainfall over a period of months. At the same time, the flooding on the Missouri bore witness, as has other severe flooding worldwide, to the fact that large dams do not provide ironclad insurance against flooding and may even contribute to its severity because of river bottom scouring and overtopping of reservoirs. Human beings may think they can control and dominate nature, but ultimately they often fail to do so, with dire consequences, Cook-Lynn suggests.

The Legacy of Aswan

The writings of Nubian authors Idris Ali (1940–2010) and Haggag Hassan Oddoul (b. 1944) about the terrible social and economic costs of the displacement of the Nubians because of dams on the Nile were discussed in chapter 3. These works generally capture the dams' environmental costs in indirect terms, by allusions to their effects on the Nubian population that remained in the Nile basin. In *Dongola: A Novel of Nubia* (1998), the childhood experience of Ali's protagonist, Awad Shalali, is painted in grim colors: "Drought and famine. It was his bad luck to have been born in the time of famine. He remembered his miserable childhood with anguish, spending nights with an empty stomach when they had not succeeded in finding wood for cooking. He ate a disgusting thick paste called *ambudays*, made of brackish water, oil, salt, and black bread. He ate rotten pickled fish called *tarkeen*, left uneaten even by the dogs."[33] Disease is rampant: "His body was anointed with Nile mud full of bilharzia parasites. Only chance kept him from death, though epidemics killed many of his friends: malaria, cholera, yellow fever, and tuberculosis, to say nothing of scorpion stings and snake bites. It was a forgotten land, which the world had erased from its map, to serve as a reservoir for its water."[34] The dearth of wood, the brackish water, the many waterborne diseases—all are implicit and well-documented negative consequences of reservoir stagnation and pollution. The out-of-season floods linked to the careless raising and lowering of the dam reservoir had a detrimental effect on agriculture: "In summer, when the river dropped and the land was visible, he went down with his mother to mow the couch grass and plant seeds. They sowed, but only rarely reaped. The north had glory and leisure; the south, death and floods. How many times had they been ruined by a flood coming out of season?"[35] This excessive saturation of the soil had caused the mud drowning death of Awad's grandfather years before. The manipulation of the Nile waters is conducted with a minimum of thought given to its impact on the Nubian farmers. Implicitly, their grotesque sacrifices are, as usual, warranted in the cause of supposed development.

The major protagonist of Ali's *Poor* (2005) is more explicit about the links between the Aswan dams and the ruin of the land and deadly subversion of the water:

> Man, because of how young you were back in those dark days, you did not grasp what was happening and what was going to occur. There was a direct link between the flood, the dam, and bread—not to mention the dreadful locusts that sometimes appeared.
>
> The next morning and throughout the days that follow, you watch the river as it creeps toward your homes up in the fold of the mountain. You see what useful things the river brings down from the villages of the South. The youths

128   CONTAMINATED WATER, DISAPPEARING FISH, AND DEADLY SEDIMENT

who swim best race out to snatch what might be used as feed or firewood. One day, two swimmers race out to reach a shapeless black object floating with the current. The one who wins the race wrestles with the thing. After a short struggle, the two disappear into the belly of the river. It turns out that the object is a vicious, nasty crocodile.

From that day on, swimmers stop pursuing the floating objects.[36]

The expansion in potential damp breeding grounds that accompanies the formation of dam reservoirs and irrigation favors an increase in locust populations, and crocodiles worldwide have welcomed reservoir conditions in subtropical and tropical areas. Neither of these developments serves humanity well. Ali's example brings painful drama to this problem.

When the protagonist of Ali's *Poor* moves to Cairo, an explicit contrast points once again to the pernicious impact of the dams on the Nubians' agricultural production: "From the absolute poverty of Nubia to the relative poverty of Bulaq [an area of Cairo]. From the post-dam scarcity of food to the abundance of it here, at least for those who can afford it."[37] Such observations undermine the often-touted agricultural benefits of dam creation. The post-dam world carries with it an apocalyptic tinge that literature effectively captures.

The stories included in Oddoul's collection *Nights of Musk: Stories from Old Nubia* (2002) also bear witness to the environmental damage caused by the Aswan dams. In "Adila, Grandmother," the narrator mentions fetching "cloudy canal water for the jars in our houses."[38] In "The River People," polluted water, loss of good agricultural land, and problems with scorpions and snakes caused by dam-related water flow are mentioned: "At flood time after the dam was built, the water would flow over the farky [a depression in the land along the edge of the river] and surround us. The scorpions and snakes fled in its wake, coming up to the villages from inside the mountains. Incidents of stinging and death from poisoning increased"; "The water swelled up like boiling milk, and as it rose it swallowed up half the green valley and destroyed it. It drowned lines of palm trees and polluted the sweet water . . . We crowded in on the scorpions, and they crowded in on us. We chased away the snakes, and they came back and surrounded us."[39] As with the reference to locusts and crocodiles in Ali's *Poor*, an almost biblical sense is expressed of an environmental plague on the land that the dam has caused.

Contending with Threatened Wildlife in India

The fate of wildlife in areas dam reservoirs will inundate often assumes a low official priority. Goldsmith and Hildyard quote from a report on India's infamous Narmada River Project:

> Characteristically, there is no plan to relocate the wildlife [in the threatened area]. When we asked officials about this, they stated that there would be "natural relocation,—that is to say, the animals would move out of the area to be flooded by themselves." Such an assumption, however, seems to us to [be] more of a sick joke and a convenient excuse than a serious proposition. Other than birds, and possibly a few alert mammals, how many animals really stand a chance of "relocating themselves" when the waters come their way? How much adjoining forest is there, anyway, for them to move into? Will not such a movement increase competition between animals (especially the strongly territorial ones) in the new habitat?[40]

These fears prove to be highly justified. McCully notes that when wildlife rescue plans are implemented, only a small proportion of the animals are saved; most drown or starve to death. Those saved often do not survive for long.[41] Competition for resources between human beings and animals may also increase in areas near dams, causing damage, hunger, even death.

In *A Dirge for the Dammed* (1992), Vishwas Patil's award-winning saga of Indian villagers displaced by a dam discussed in chapter 3, the narrator almost immediately alludes to the environmental consequences of dams. Khairmode Guruji, the teacher who is the longtime champion and organizer of his suffering neighbors in Jambhli, has already seen—while employed in another dam-affected area—villagers forced to move higher into the hills, with no electricity and no money for kerosene, eating by firelight and washing their clothes in "stagnant clayey water."[42] In speaking with officials about the dam planned for Jambhli, Guruji brings up the suggestion increasingly offered by environmentalists of building multiple smaller dams instead of one large one: "Build three smaller dams instead of this one. Let us remain where we are. These people will get water without having to bear the burden of our resettlement. They will benefit, we will prosper, there will be development all around." Guruji's arguments fall on deaf ears: "Such nonsense he speaks. Who does he think he is, the chief engineer? The bigger the dam, the larger the water reservoir. More water means more progress. No one had the time to listen, to consider alternate proposals."[43] In fact, as the novel shows in painful detail, more water does not mean more progress but less, for those called upon to make sacrifices.

The Jambhlikars have to move when the site of their village is entirely submerged. A few miles away, the village of Jhanjadwadi, where some Jambhlikars have relatives, remains, but it still suffers environmental consequences. An initial incident marked by Patil's characteristically graphic sensational manner captures the danger now afforded by desperate wild animals:

> Two days later they got the news that Farshya Kathodi had been attacked by a bear. The wadi [valley] paled with fear. Everyone rushed to Farshya's hut. They found him lying on the ground, naked except for a piece of cloth around his waist. His body was raked by claw marks, as if a thorny comb had been drawn over him. He was bruised all over, his ballooning back, black and blue. His face was badly mauled. The bear had lived up to its reputation . . . The wadikars were shaken by the incident. Gundya Malusare was the first to speak, "I had heard that bears were flourishing in the forests of Irlewadi. They must have spread to ours as well."
>
> "The dam has sealed their routes. Naturally they will multiply here. In the evenings as you walk down, you can barely avoid stepping on young snakelets—they are all over the place," Gunvanta replied.[44]

Soon there is an incident in which hyenas kill sheep from someone's flock. When one villager expresses surprise that hyenas are killing live animals rather than scavenging for dead ones, the sheep's owner replies: "There's no village left on the other side now, so where will it [the hyena] find dead animals? When they can't find anything else, they have no choice but to gnaw at my bones, and those of my flock."[45] Both animals and humans now suffer because of the unnatural impacts of the dam.

Animals soon multiply, unchecked, and seek food wherever they can find it. Wild boars repeatedly attack crops, and the villagers cannot afford enough ammunition to keep them at bay. Hyenas maul sheep in their pens. Prowling bears are a serious problem. Dangerous snakes slip into homes. Insects and reptiles multiply and lurk underfoot, to the point that villagers are afraid to walk the forest paths alone. Tigers attack cattle and drag away lambs and calves. Minimal official assistance is provided to deal with these problems.

Once again, predictable environmental damage is ignored, while both animals and disadvantaged human beings suffer and are set at greater odds with one another.

Sedimentation and Earthquakes on the Yangtze

Of all the environmental consequences of large dams treated in the literary works discussed in this chapter, reservoir sedimentation may appear to be the least sensational because of its comparative invisibility; in terms of ruinous impact, however, it is one of the worst problems and a development whose seriousness has become increasingly apparent over the past few decades. Earthquakes readily evoke apprehension, but the potential for earthquakes near dam sites was long unrecognized or dismissed. That is no longer the case, and there is growing awareness that a variety of circumstances may trigger dam-related earthquakes.

As noted above, reservoir sedimentation is inevitable, a kind of planned obsolescence but one that little concerned early dam builders. McCully declares that "a river, in effect, can be considered a body of flowing sediments as much as one of flowing water."[46] By their very nature, dams prevent silt and other solid substances from flowing downstream. How much silt accumulates behind a dam is hugely affected by soil erosion, which, in turn, is driven by deforestation. Such erosion may take place far upstream from a dam, on a river's tributaries, for example. Soil erosion due to deforestation is a particular problem in the tropics: "The organically poor soils of the tropics are particularly vulnerable to erosion, and although the monsoons last for only a short time, they can quickly wash away the soils from deforested slopes."[47] While the need for reforestation is often recognized, it is frequently given low priority, particularly in the interests of short-term financial benefits from dams and land clearance. Even when some reforestation takes place, the trees planted are not necessarily the most appropriate for soil retention. Neither are the muddy wastelands that develop behind a dam as a result of premature sedimentation suitable for agricultural usage; the soil is too tightly compacted. In addition, the sediment may exhibit a pronounced accumulation of pollutants. Downstream, as mentioned, the loss of nutrient-rich silt has a multitude of negative results. Finally, as should be obvious, dam failure due to sedimentation means a loss of the hydroelectric

132 CONTAMINATED WATER, DISAPPEARING FISH, AND DEADLY SEDIMENT

capacity that was often a major reason for the construction of the dam in the first place. Hence the current worries about the lowering of Lake Mead, for example.

It is "extremely difficult to estimate how much sediment will be trapped by a reservoir," and sedimentation of reservoirs can occur surprisingly quickly.[48] Most reservoirs "trap nearly 100 percent of the river sediment loads that enter them"; the threat of significant sedimentation of reservoirs is exacerbated by increasing erosion driven by changing land use methods and intensity and is compounded by a changing climate.[49] The rate of sedimentation is not constant; floods may dramatically increase the amount of sediment a river carries. Even one large instance of annual precipitation can cause a major increase in sediment volume.[50] No truly effective solutions to the problem of sedimentation have yet been discovered: "There are three categories of methods to prolong the life of a reservoir: reduce the amount of sediment flowing into it; flush through the dam the sediment that has already accumulated; or dredge the sediment. All have severe limitations, either because they simply do not work, they are prohibitively expensive, or because they conflict with the dam's ability to supply water and power."[51] This is a striking example of ill-considered long-term planning in favor of short-term gains.

The threat of reservoir-induced seismicity—that is, earthquakes—is extremely fraught. The first recognized instance of such activity occurred in the 1930s, in connection with Lake Mead, the reservoir behind the Hoover Dam.[52] For several decades, though, such earthquakes were the subject of much debate and often regarded as freak occurrences. Only since the late 1960s has the problem been generally acknowledged. The conditions under which reservoir-induced earthquakes may occur are still not completely clear. While it was originally believed that the threat existed "only while a reservoir was being filled, or immediately after it reached its maximum height," it now appears that earthquakes can occur when the water level in a reservoir is being lowered, when it has been emptied and refilled, or even when the water level is relatively stable.[53] Such earthquakes can happen in areas known for their seismic activity and in areas considered seismically inactive.[54] In other words, the potential for earthquakes in areas near dams is much greater than was acknowledged even a few decades ago.

Long before the Three Gorges Dam was completed, experts issued warnings about the dangers of sedimentation that would be associated with it.

Contaminated Water, Disappearing Fish, and Deadly Sediment   133

Such warnings figured prominently in the provocative collections of essays assembled in the late 1980s and 1990s by the then imprisoned Chinese journalist Dai Qing mentioned in chapter 3. For example, an extensive report prepared in 1996 for the United States Export-Import Bank, which was considering providing financing for the dam, concluded that because of anticipated sedimentation problems, "investing in the project would be unwise."[55] Such typical understated bureaucratese conceals acknowledgment of a potentially serious problem. The bank decided not to support construction of the Three Gorges Dam. As for reservoir-induced seismicity and Three Gorges, a growing number of scientists believe the dam has contributed to earthquakes—some horrendous—in the region in recent years.[56] Such concerns and fears underlie Dan Armstrong's *Taming the Dragon* (2007) and Alex Kuo's *The Man Who Dammed the Yangtze* (2011). Both novels are noteworthy for their skillful and extensive incorporation of scientific and mathematical detail.

Armstrong is a novelist with a degree in aerospace engineering and the owner and operator of the website Mud City Press, an online magazine devoted to the environment and sustainable agriculture. In the acknowledgments to *Taming the Dragon*, Armstrong credits another, earlier American author for inspiring the writing of his own novel: "I also owe a debt of insight, imagery, and inspiration to John Hersey. He wrote this same story fifty years ago in his novella *A Single Pebble*."[57]

John Hersey (1914–93) was a prolific writer and journalist. He is perhaps best known for his lengthy piece "Hiroshima" (1946), an account of the bombing that draws heavily on his interviews with six survivors of the catastrophe.[58] Hersey was born in China, the son of Protestant missionaries. His family returned to the United States when he was ten. Hersey went back to China briefly as a journalist in the late 1930s and made a trip down the Yangtze in 1946. *A Single Pebble* (1956) is set in the 1920s. The first-person narrator, a naive and ambitious young American hydraulic engineer, travels up the Yangtze on a junk, enthralled by fantasies of constructing a huge dam on the river. From the beginning of the novel he is explicit about his dreams and the impetus for them: "Being an ambitious young engineer I could only think of it [the Yangtze] as an enormous sinew, a long strip of raw, naked, cruel power waiting to be tamed."[59] This is very much in keeping with the kind of high modernist dreams of conquering nature discussed in chapter 2.

At Witches' Mountain [Wu] Gorge, one of the famous Three Gorges, the narrator believes he has identified the perfect site for a gigantic hydroelectric dam: "There it was! Between those two sheer cliffs that tightened the gorge a half-mile upstream, there leaped up in my imagination a beautiful concrete straight-gravity dam which raised the upstream water five hundred feet . . . Away through pipelines flowed, too, unimaginable numbers of acre-feet of water, irrigating lands that after the harvest would feed, let me say, seventy-five million Chinese. A terrible annual flood . . . was leashed in advance by this beautiful arc. Beyond the tall barrier, junks sailed forward with their wares, to Chungking and farther, as on a placid lake."[60] When the narrator excitedly attempts to share his wondrous vision of human-inspired agricultural and industrial wealth with the Chinese owners and trackers on the junk, they respond with hostility and deliberately exclude him from their conversation. Their reaction and apparent lack of interest shock the narrator.

As the junk progresses through the Three Gorges, however, he begins to doubt his own commitment to technological progress above all and to wonder about the possible merits of the implicit Chinese attachment to the Yangtze's natural regime: "For me the unendurable idea forcing itself upward . . . was that perhaps Old Pebble [the head tracker] was right: perhaps the Great River could not and should not be challenged by such as I; perhaps a millennium-in-a-day was after all not something that could be *bestowed*."[61] The narrator's doubts subside, however, and four months later, after further study and another trip downriver, he submits his recommendations, but to no avail: "I wrote an optimistic, even fervent, report on the possibilities of a dam in Yellow Cat Gorge [Xiling Gorge, the eventual site of the Three Gorges Dam], where . . . the site seemed to me best of all. It is clear that nothing ever came of that report, or of me . . . The dam is still to be built. It will be, one day—of that I am sure."[62] And it was, with dire consequences.

A Single Pebble is marked by a disquieting sense of ambivalence. There are overtones of cultural prejudice in the narrator's interpretation of the Chinese view of a dam as nothing more than an unfortunate event rather than a crime against nature. In his study of the Yangtze, Lyman P. Van Slyke observes, "For all his newfound humility, Hersey's engineer patronizes the Chinese still by assuming—indeed somehow admiring—a preference for patient endurance rather than for change."[63] As Van Slyke also points out, many Chinese were themselves high modernist devotees. Consider Mao's

Contaminated Water, Disappearing Fish, and Deadly Sediment 135

poem about the Yangtze. Yet, despite the narrator's eventual recommendation of a dam, the cumulative impression of *A Single Pebble* is that technological progress should not be an uncritically embraced goal.

Taming the Dragon also questions the wisdom of constructing a mega-dam on the Yangtze, but it does so much more on explicit environmental rather than cultural grounds. The major protagonist of the novel is Hans Fruehauf, a young German hydrologic engineer with a fluent command of Mandarin who works for a German bank as a technical adviser for its Asian investments. Hans comes to Asia with great expectations for the Three Gorges Dam, which is now near completion. His optimism is shaken, though, even before he arrives at the dam site. The novel's opening line is spoken by one of two North American engineers at a Singapore hotel, where Hans is staying on his way to China: "What a fiasco! The world's largest dam. Ha! The Chinese are out of their minds."[64] When Hans suggests that only minor construction details need to be worked out at the dam being built at Xiling Gorge, the engineers provide an explanation for their obvious disbelief and contempt. First and foremost is the issue of silt: "If the dam doesn't simply fail, my guess is that the reservoir will silt up before all those minor details are worked out."[65] The other North American offers an unfortunate comparison: "In my eyes the dam is no more than a great lump of concrete, vastly less aesthetic and only slightly more practical than the Great Pyramid of Cheops."[66] Such an invidious comparison with the Egyptian pyramid is the precise opposite of the usual comparisons of large hydroelectric dams with pyramids. Another negative comparison is soon forthcoming. When one of the engineers asks another engineer for his opinion of the Three Gorges Dam, the latter replies, "It will make the Aswan look like one of the Seven Wonders of the World."[67] Given the widely recognized problems that resulted from the construction of the Aswan High Dam, this observation has more than a slight sarcastic tinge.

At the conclusion of the conversation, one of the engineers provides Hans with a critical assessment of both the very idea of the dam and its political origins: "If they'd done it right, they'd have built eight cofferdams on the various tributaries upstream of the site They'd get the same amount of electricity with greater control of the flood plain and considerably less impact on the environment. Man, it's just a huge political snow job. Hype for the largest dam in the world. The crazy dream of some ambitious party

136 CONTAMINATED WATER, DISAPPEARING FISH, AND DEADLY SEDIMENT

chairman who's been dead thirty years now."[68] Chinese opponents of the dam had, in fact, long argued for multiple small dams, but their suggestions were resisted.[69] Once again, as elsewhere, bigger seems better. Despite the concerns he hears during his Singapore stay, Hans still hopes for the best.

When Hans arrives in Wuhan, he has an initial conversation with three Chinese engineers, whom he questions about silt and the issue of several smaller dams versus a single large one. The Chinese by no means completely dismiss these issues but ultimately emphasize their country's political and developmental concerns and the need for compromise: "There is a national pride and a politics in this dam that also enters into the calculus of our decisions. China wants to industrialize. We want energy. We want control of the Great River. We want a deeper shipping channel upriver to Chongqing. These things only come with compromise. Minimize this. Maximize that. That is what engineering at this level really is. A difficult progression of trade-offs."[70] In other words, compromise, and don't worry about the consequences. Hans is quick to reassure his interlocutors that he does not mean to sound critical and is, in fact, very optimistic about the dam's success.

When Hans first sees the dam construction site, he is initially naively impressed, "swelled with pride to be part of such an epic undertaking."[71] Closer inspection offers a series of jolting disappointments. The first involves silt and its dangers. The dam reservoir is much lower than expected, and the water is "a disturbing coffee with cream color,"[72] an effect of excessive silt. It also exudes an unpleasant odor. Hans is keenly aware that his bank is hoping for and expecting a positive report and initially tries to assuage his own concerns, but doing so becomes increasingly difficult. The sight of the reservoir calls to mind reports Hans has read. Produced before the construction of the Three Gorges Dam began, the reports suggested that high dams should no longer be built because of the environmental damage they cause. He also recalls the criticisms human rights activists have expressed about the scale of the intended relocation of people. For the first time, Hans begins to harbor doubts about the undeniable benefits of hydroelectricity. Like Hersey's narrator in *A Single Pebble*, the young German engineer begins to question his own unadulterated commitment to technology: "Little things seemed to stand out with [the] ever-present suggestion that there was something more to know here in China than hydrology."[73] And above all, there is the silt. As Hans watches barges furiously dredging near the base of the dam, he

Contaminated Water, Disappearing Fish, and Deadly Sediment 137

wonders seriously for the first time about the security of his bank's investment. Silt is an enemy that can scarcely be defeated.

Worries about pollution are soon added to Hans's concerns about silt. A one-on-one conversation on the dam walkway with the engineer Huang Shanchang brings these concerns to the foreground:

> [Huang Shanchang] turned away and, in an exaggerated way, lifted his head to breathe the air. "The air is sweet," he said.
>
> The smell of the reservoir was anything but sweet, thought Hans. "How so?"
>
> "Within the thick scent of sewage is a sweetness," said Huang Shangchang . . .
>
> "Do I misunderstand your word *sweet*" [asks Hans]?
>
> "Like heavy metal. Arsenic, cyanide, methyl mercury."
>
> Hans breathed in again. Oh, yes. Threading through the thick fecal smell of sewage was a sharp metallic sweetness. "What is this?"
>
> "Chemicals leached from hundreds of submerged factories and manufacturing plants upriver."[74]

Implicitly, Huang Shangchang cannot speak openly to the authorities about such problems. Hans still clings to the hope that the dam's benefits to China will offset the obvious drawbacks, but Huang Shangchang provides no reassurance, comparing the river to an artery in the human body. Hans concludes, "The artery is blocked and its blood is poisoned."[75] Ultimately, Three Gorges promises death, not wonder-filled life.

Hans decides that to come to an informed conclusion about the Three Gorges Dam, he needs to travel upriver to the city of Chongqing at the western end of the reservoir. There, he intends to do further investigation into sedimentation while scuba diving. The boat trip across the reservoir affords Hans more glimpses of many kinds of pollution: "Styrofoam cups, plastic bags, baby diapers, and paper wrappers were common flotsam on the water surface. Throw in the raw and invisible sewage. Progress, it seemed, had made a monstrous oriental stew of the Yangtze."[76] This might appear to be an unwontedly sensationalistic assessment of pollution on the Yangtze, but unfortunately it conforms to reality. At the beginning of *Before the Deluge* (2002), Deirdre Chetham provides a firsthand account of the pollution she witnessed on a cruise in 1983:

138 CONTAMINATED WATER, DISAPPEARING FISH, AND DEADLY SEDIMENT

Outside the dining room windows, fishermen and vegetable merchants in rowboats and sampans floated by, along with bits of rubble and tin cans, parts of houses, and drowned livestock. Anything not needed or wanted, from leftover food and garbage to large pieces of machinery and rattan couches, was tossed overboard or off the riverbank. *"Gei Changjiang chi!"* (Give it to the Yangtze to eat), someone would say, heaving the object into the water from the shore or boat. Humans sometimes met the same fate. While serving dinner, the waitresses would cry *"siren, siren"* (corpse, corpse) every time a bloated, decaying body of a suicide, a villager too poor to be buried, or a newborn baby girl drifted by. The dinner staff would quickly analyze the cause of death and then avert their eyes, announcing that it was bad luck to stare at the dead. Local officials reportedly felt it was equally bad luck to deal with these *shui do bang*, or water logs, as they were called, and rumor had it that corpses were only picked up if there was anything left of them by the time they reached Shanghai.[77]

Like the Chinese fellow trackers and boat owners in *A Single Pebble*, the riverboat captain in *Taming the Dragon* tells Hans that he sees things differently than the dam builders and warns: "Man cannot control the Great River."[78] Here, the captain is implicitly calling into question the high modernist conviction, now fully embraced by the Chinese establishment, that human beings can triumph over nature. In this the captain echoes the perspective afforded at times in *a Single Pebble*: that nature should be left to its own devices.

At a stop on the way to Chongqing in the town of Fengjie, Hans goes for a dive and collects water samples and plant and animal specimens, such as small fish and worms. The below-water sight is disturbing—broken structures, silt, no vegetation on the river bottom, algae, scavenger fish with ulcerated flesh. The samples and specimens confirm the presence of dangerous levels of industrial chemicals and organic waste. Hans concludes that "if a monumental cleanup did not begin soon, a huge environmental disaster was in the making—if not already made."[79] Then, presumably because of his excessive immersion in the filthy water, Hans sinks into illness marked by stomach distress, fever, and chills. There is no hint in the novel that officials have any cleanup in mind.

After further adventures in the hinterland with modern-day Chinese gangsters and the start of a torrid love affair, Hans returns downstream and has

Contaminated Water, Disappearing Fish, and Deadly Sediment 139

an opportunity to compose and deliver a report to senior bank employees. Hans's report explicitly outlines the catastrophic nature of the Three Gorges Dam project and suggests that the bank invest no more money in it until the reservoir is cleaned up. His most senior colleague refuses to accept his conclusions and demands that he rewrite the report. Hans tries to be accommodating, but ultimately, at a meeting of the bank representatives with Chinese authorities, he cannot restrain himself and is passionately explicit about his environmental concerns. The Chinese react with hostility; Hans's colleague reacts with anger and eventually declares that Hans is fired. The young hydrologist announces that he will return to Berlin to discuss the situation with the bank directors, saying "personally, I don't believe [the bank] is about investing in toxic waste dumps."[80] Hans is well aware that the Three Gorges Dam project will most likely still proceed to completion, but he has been true to his conscience. And he has won the woman of his dreams, with whom he plans to fly to Paris. The reader may admire Hans's honesty but nonetheless recognize its inefficacy.

Armstrong hoped that *Taming the Dragon* would "share some kinship with Hersey's fine novella."[81] In that, it succeeds. Like *A Single Pebble*, *Taming the Dragon* tracks the broadening cultural consciousness of a young hydrologic engineer and offers a critique of the unquestioned benefits of technology. Where *A Single Pebble* is subtle, *Taming the Dragon* is devastatingly explicit. Like Gulick in *The Dam Builders* but even more effectively, Armstrong presents a detailed catalog of ecological issues in an accessible and highly engaging way. *Taming the Dragon* demonstrates that contemporary novels about the environmental consequences of large dams may now draw upon a huge body of research by scientists and warnings by activists to make a compelling case for concerns about mega-dams.

Alex Kuo (b. 1939) is an author and retired professor of English who has received many grants and awards and who lived in the Columbia River Basin for many years. His novel *The Man Who Dammed the Yangtze* tackles some of the same topics as *Taming the Dragon* but with a touch of magical realism. The novel's two major protagonists are G (short for Gregor), an Asian American mathematician and academic, and Ge, a female Chinese mathematician and academic. The novel recounts their parallel careers across several decades. It begins in 1968, a year in which G experiences the political and social upheaval taking place on North American college campuses because

140   CONTAMINATED WATER, DISAPPEARING FISH, AND DEADLY SEDIMENT

of the Vietnam War, while Ge observes firsthand the depredations of the Red Guard during the horrific Cultural Revolution. G first learns about the plans for and background to the construction of the Three Gorges Dam from Amy, a radical young faculty member who gives him "a short course on the men who dammed the Yangtze" at a cocktail party.[82] Both G and Ge soon decide to quit teaching—G because he is disgusted by his department's support for the police crackdown on a campus protest mounted by African American students, Ge because she has lost her sense of purpose in teaching. G takes a job with Westinghouse in Pittsburgh doing applied mathematical calculations, Ge a position in the engineering department of the Yangtze Valley Planning Office in Chengdu. The two have still not met.

In 1972, Ge is asked to prepare projections of the siltation problem for the proposed Three Gorges Dam. When she asks her supervisor, Zhang, for a couple of days off, supposedly to visit temples in nearby Dujiangyang, Zhang suggests that while she is there, Ge visit the city's ancient irrigation system. Built more than 2,000 years ago during the Qing Dynasty but still functioning beautifully, the irrigation works at Dujiangyang are a UNESCO (United Nations Educational, Scientific, and Cultural Organization) World Heritage Site. As is explained in *The Man Who Dammed the Yangtze*, the three main water gates of the Dujianyang project "controlled the distribution canals diverting a good portion of the fast-flowing Min River into a year-round, intricate irrigation network of succeedingly smaller and more complex system of canals, weirs and sluices that irrigated three million hectares of agricultural land . . . the Flying Sand Spillway at the head of these gates . . . prevented flooding in the late spring and summer months when the run-off from the steep Qingcheng Shan [a mountain, also a UNESCO World Heritage Site] appeared ominous."[83] This example points to the fact that newer ideas about damming are not necessarily superior to old ideas and gives a lie to the idea that one gigantic dam is superior to a series of more modest water control constructions.

In Dujiangyan, Ge meets Weng, an older man whose family has been monitoring the dam there for more than a century. From Zhang, Weng learns that, ominously, Ge has not been given access to siltation calculations for the Gezhouba Dam being built a short distance from the intended site for the Three Gorges Dam. Ge tells Weng that she has not even been allowed to visit the Three Gorges Dam. The two discuss the amazing efficacy of the

Contaminated Water, Disappearing Fish, and Deadly Sediment 141

Dujiangyan irrigation system. Implicitly, this ancient system is far superior environmentally to the proposed new dam, and Ge's inability to visit Three Gorges or examine statistics about Gezhouba points to the victory of political over environmental concerns.

A central chapter of *The Man Who Dammed the Yangtze* is titled "Sedimentation" and contains a comprehensive overview of the Yangtze's propensity to flooding and the potential negative consequences of a large dam. In addition to sedimentation, the threat to wildlife, the submergence of more than 1,000 ancient cultural sites, water-borne diseases, displacement of people, and vulnerability to terrorism are mentioned. Ge realizes, however, that "for now, sedimentation was all [she] could work on, and she was still determined to do it right."[84] The novel includes detailed mathematical calculations and discussion of the many uncertainties associated with determining the sedimentary impact of large dams. By 1990, Ge concludes that "this Three Gorges Dam was a gigantic experiment in vanity and greed, if it worked," but that her analysis and recommendations will have no impact: "One way or another, the dam will be constructed, another monster tribute, like the Great Wall or the Grand Canal or the Pyramids or the Grand Coulee."[85] It is telling that the pyramids and Grand Coulee are not contrasted here but are lumped together as monsters.

This is by no means the first or the last reference to Grand Coulee in *The Man Who Dammed the Yangtze*. While Ge has been studying the Three Gorges Dam project, G, Amy, and Ted—G's cousin and a disenchanted veteran of the Vietnam War—have been looking at the environmental impact of North American dams. They are very much aware of the losses suffered by Native Americans in connection with large dams. At one point G observes: "The Lakotas gave up land for the Fort Randall Dam . . . the Confederated Colvilles for the Grand Coulee Dam, the Umatillas and Warm Springs and Yakamas surrendered Celilo Falls for the John Day and the Dalles Dams."[86] In 1994, G takes a road trip and visits the Grand Coulee Dam. In a chapter titled "The Robo River," G and Amy discuss and exchange extensive data about the dams on the Columbia River and its tributaries, data that include a detailed comparison of Grand Coulee with the proposed Three Gorges Dam. G begins to fantasize about the destruction of Grand Coulee: "He wondered if there was anything, anything at all, that could be done to reverse this process that has turned the environment

142 CONTAMINATED WATER, DISAPPEARING FISH, AND DEADLY SEDIMENT

and the lives of people to shit."[87] This is obviously not an endorsement for big dams.

An element of the fantastic, possible only in a literary work, now takes a prominent role in the narrative. G begins to do research on how to blow up Grand Coulee. Detailed calculations about explosives are provided. The imaginary destruction of the dam—which is presented as an actual event, although, of course, it did not really happen—is described from Ge's point of view. With colleagues in China, she watches a CNN Special Report and learns that authorities estimate "that more than a trillion cubic yards of water from Lake Roosevelt [the reservoir behind Grand Coulee] was released by the collapse of the Grand Coulee, and the force of the rampaging water tore down the next dam along the river, the Chief Joseph. A wide bend in the river spilled most of the water westward into the plains, and the next dam, the Wells, held fast."[88] The architects of this supposed act of ecoterrorism are unknown, and no more is said about it in the novel.

At the end of *The Man Who Dammed the Yangtze*, G and Ge finally meet. At an international conference in Prague, they deliver papers with identical titles and the same notations and terminology. As they deliver their papers simultaneously, the dividing door between the rooms in which they are speaking retracts. The narrative ends with the two mathematicians regarding one another: "They are looking into the future, and into the past. And at this moment of moments they take a breath and look at each other, the audience, the author and the reader."[89] It is clear that G and Ge have similar visions of dam damage but equally clear that their warnings will be ignored.

With its fantastic twists, *The Man Who Dammed the Yangtze* offers a unique perspective on the environmental consequences of large dams. The problem of sedimentation—which, as experts have repeatedly noted, is inevitable, dangerous, and costly—is given pride of place in the narrative. Ge's calculations and realization that Chinese authorities will ignore her analysis bear witness to the political preoccupations that have long marked the construction of large dams throughout the world. G's imaginary destruction of Grand Coulee points to a possible, if extreme, solution to this fraught dilemma. This is fictional environmental criticism in its most dramatic form.

The impact of large dam construction on peoples forced to relocate from dam sites was immediately recognized, if imperfectly and incompletely understood or deliberately minimized. The environmental consequences of large dams, however, were initially underestimated and continue to be exposed. There is now a widespread appreciation of the ecological harm done by large dams and the vast pollution that has overtaken many of the world's rivers. In *Don't Cry, Tai Lake* (2012), a murder mystery by the American author Qiu Xiaolong that contains extensive discussion of water pollution and official corruption, the following conversation takes place between two characters: " 'Talking about Mao, do you remember the picture of Mao swimming in the Yangtze River?' . . . 'Well, with China's rivers and lakes so polluted now, Mao jumping into the river would be seen as a suicide attempt.' "[90] The comment is sarcastic and underscores the Yangtze's pollution. This conversation bears witness to the shift in thinking about potential human control of waterways that has taken place over the past century.

The novels and stories treated in this chapter span almost fifty years. Over time, their focus has expanded from consideration of obvious and immediate detrimental environmental effects to include longer-term, but no less dangerous, ecological impacts. The submerged and rotting trees, brackish water, and decline in edible fish populations Rasputin describes are now complemented by Ali's disease-infested mud, Patil's violent confrontations between stressed animal and human populations, and Kuo's detailed attention to the buildup of sediment in reservoirs and its pernicious consequences.

As is readily apparent, the writers who treat such topics are increasingly more highly informed about the manifold varieties of environmental damage linked to large dams. The level of scientific detail that underpins Armstrong's *Taming the Dragon*, for example, is now possible because of the many works—some intended for a general audience—that have been produced by a wide spectrum of experts and activists in recent decades. To their great credit, a number of authors of fiction have demonstrated a remarkable ability to present sophisticated material in an accessible and compelling way.

As is also perhaps obvious, receptivity to literary exposés of the environmental consequences of large dams is not uniform worldwide. While North American, Egyptian, and Indian authors have been able to produce and have published in their own countries devastating fictional critiques of a number of real and imaginary dams, negative representations of the Three Gorges Dam, for example, have been limited to foreign authors—many of them Asian Americans with vast experience in China but able to work with Western publishers only because of their expatriate status. Will this situation change? Perhaps not soon, but Rasputin's publication in Russia of a story like "Downstream and Upstream" provides an indication that official government receptivity to criticism need not remain static. The rising tide worldwide of environmental concern should encourage and facilitate the writing and publication of more novels and stories like those discussed in this chapter, as well as increase popular interest in this crucial topic.

5

Dam Failures, Real, Imagined, and Ecotage-Inspired

Dams occasionally fail; even more frequently than they fail, people are terrified by the threat of dam collapse. Rapid deluges that result from dam failures are frightening, dangerous, and potentially lethal. Dams may give way for many reasons—poor construction, inadequate maintenance, overtopping and ensuing dam collapse caused by heavy rains, geological insecurities, earthquakes, deliberate destruction, or terrorism. There have been dam failures for thousands of years. For example, the earthen Great Marib Dam, located in present-day Yemen and built in the eighth century BCE, was definitively breached in the sixth century CE, despite repeated repairs over more than a millennium. Maintenance may have been an issue, but there are also possibly apocryphal tales of responsibility due to large rats gnawing at the foundations. The number of flood victims of the Great Marib collapse is unknown, but the dam's impressive ruins are still visible, as is its modern replacement.

In the United States, the worst dam failure to date has been the Johnstown Flood that occurred in Pennsylvania on May 31, 1889, when the South Fork

https://doi.org/10.5876/9781646425976.c005

Dam in the heart of the Alleghenies gave way. Yet although more than 2,000 people perished and a national memorial marks the site of the catastrophe, surprisingly few people have heard of the Johnstown Flood (this observation is based on my own anecdotal evidence).[1]

A few decades later, on March 12, 1928, the St. Francis Dam in San Francisquito Canyon near Los Angeles collapsed.[2] There had been signs of incipient problems on the day of the collapse, but the dam's designers, who included most prominently the self-taught and highly successful civil engineer William Mulholland (1855–1935), took a supposedly careful and close look at the construction and declared that there were no reasons for concern. Nearly 500 people died. Even later in the twentieth century, the Teton Dam in Idaho collapsed, largely because of known but dismissed geological site problems. Thankfully, because of a highly successful warning system, only 11 people died when the dam collapsed, but many thousands of cattle perished.

India, Brazil, Italy, China, and Libya have also been the locations of terrible dam failures. On August 19, 1917, the Tigra Dam, located in Gwalior in the central Indian state of Madhya Pradesh, collapsed when water infiltrated the dam's foundation. At least 1,000 people died. On July 7, 1961, the Panshet Dam near the western Indian city of Pune gave way because of heavy rain and shoddy construction. Again, at least 1,000 people died. On October 11, 1979, the Machchu-2 Dam, sited near the town of Morbi in Gujarat on the west coast of India, failed, again in part because of extraordinarily heavy rains. At least 5,000 people died.

In Europe, on October 9, 1963, Italy's Vajont Dam, located near the northern mountain of Monte Toc, was overtopped because of a massive landslide in the dam's reservoir; in an all too familiar scenario, engineering assessments before construction had been faulty. More than 2,000 people died. Even more recently, on January 23, 2019, the Brumadinho Dam in Minas Gerais, a southeastern Brazilian state, failed, leading to the deaths of 270 people. The most staggeringly horrific recent catastrophe in the past century, however, was the previously mentioned collapse of China's Banqiao Dam, on the River Ru in Henan Province, on August 8, 1975. In a cascading domino effect, sixty other dams gave way. Design flaws and massive rains at the time from Typhoon Nina contributed to the disaster, because of which as many as 240,000 people died and 11 million people lost their homes. The accuracy of

Dam Failures, Real, Imagined, and Ecotage-Inspired   147

these numbers is by no means certain; they may well be higher since China often cloaks its statistics in a veil of secrecy.

As for deliberate destruction of dams, mention should be made of Operation Chastise, which the 617 Squadron RAF Bomber Command conducted on the night of May 16, 1943.[3] The highly imaginative engineer Barnes Wallis designed an unusual type of "bouncing bomb" that was able to breach the Möhne and Edersee Dams in Nazi Germany's industrial Ruhr district. The damage was catastrophic but did not meet the hoped-for expectations of crippling German industry; of the 1,600 deaths that occurred, almost 1,000 were mainly enslaved Soviet laborers. Operation Chastise did, however, give a major boost to British morale, and in 1955 the war film *The Dam Busters* gained lasting fame as an example of courageous Allied efforts to defeat the Germans.

Dam disasters, real and imagined, have figured largely not only in film but also in fiction, from the trite and trashy to the truly moving and frightening. This chapter examines a sample of such works by a variety of authors, largely American, often playing on the tensions of the Cold War or the threat of Middle Eastern terrorist groups. The increasing popularity of such writings provides testimony to a growing awareness of the dangers mega-dams may present—as horrific examples of the liabilities various groups and even nations may exhibit because of complex problems of poor construction and scientific ignorance—and the gripping manner in which literature may represent such dangers. The result of such awareness, from whatever source, may in fact be environmental sabotage intended to restrain the growing impact of humanity on nature and the concomitant desire to dominate and control natural phenomena.

The Johnstown Flood

In recent years, the Johnstown Flood has become the subject of an increasing number of novels and interactive games, particularly for juveniles. Examples include John and Lisa Mullarkey's *The Johnstown Flood: An UP2U [Up to You] Historical Fiction Adventure* (2014) and Stephen Otfinoski's *Can You Survive the Johnstown Flood: An Interactive History Adventure* (2022). The two novels about the Johnstown catastrophe that I have selected for discussion are Kathleen Cambor's *In Sunlight, in a Beautiful Garden* (2001) and Judith Redline Coopey's

148   DAM FAILURES, REAL, IMAGINED, AND ECOTAGE-INSPIRED

Waterproof: A Novel of the Johnstown Flood (2011). Cambor's novel deals largely with the period before the disaster, Coopey's with the aftermath. Both novels effectively convey a sense of the dread and horror that marked the events that took place in Johnstown, Pennsylvania, in 1889 and gained worldwide attention.

Ironically, the earthen South Fork Dam, about twenty-four kilometers north in the hills above Johnstown, was initially well constructed and apparently safely designed in the first half of the nineteenth century.[4] Its reservoir covered 450 acres and was almost twenty-two meters deep in places. Intended as part of an extensive Pennsylvania canal system, the dam, however, was quickly abandoned because of the advent of the railroads, which almost immediately made most canals obsolete and unnecessary. Virtually untended for more than twenty years, in 1879 the South Fork Dam was purchased by Benjamin Ruff, an entrepreneur and real estate dealer, for recreational purposes for the wealthy. Understandably, the dam had suffered great deterioration over the decades. For years there was constant talk in and outside of Johnstown about the possibility of the dam breaking, but there were equally frequent assertions that there was no cause for alarm. Daniel J. Morrell, a prominent Johnstown businessperson (Johnstown was the site of the town's largest employer, the Cambria Iron Company), sent the experienced engineer John Fulton to assess the dam. Fulton asserted in writing in no uncertain terms that the repairs to the dam in the 1870s had not been done carefully enough, but the arrogant and ignorant Ruff dismissed Fulton's concerns in cavalier fashion. Building a site of water-driven pleasure for a few rich men and their families was more important than the safety and security of Johnstown's inhabitants.

Among other problems, the South Fork Dam sagged in the middle and its height was lowered to accommodate—purely for the convenience of members of the South Fork Fishing and Hunting Club—a larger road across the dam that made provisioning the club easier. The water in the reservoir was raised nearly to the brim, again for the sailing pleasure of the club's members. The combination of the lower dam and higher reservoir proved deadly. Several inches of rain fell the day before the dam collapsed. To the general shock and horror of witnesses, it took less than an hour for the reservoir lake to empty when the dam was overtopped and collapsed. The deluge was briefly slowed behind a first bridge, which then also gave way. A second

Dam Failures, Real, Imagined, and Ecotage-Inspired 149

bridge held, but debris and people were trapped in it; for reasons that are not completely clear, a fire began. The flood's death toll might have been greater if the second bridge had also collapsed, but this was of little consolation as people heard and remembered for life the screams of the trapped and burning victims at the second bridge.

The Johnstown Flood story is one of greed, negligence, and gross sense of elite privilege. Many of the members of the South Fork Fishing and Hunting Club were prominent and highly successful Pittsburgh businessmen—men like Andrew Carnegie, Henry Clay Frick, and Andrew W. Mellon. It is possible that many of the club members knew almost as little about the structural integrity of the dam as did the inhabitants of Johnstown, but there were strong suspicions that not everything had been attended to with the utmost care. After the flood, damage suits were brought against the club by Johnstown citizens, but "not a nickel was ever collected,"[5] perhaps in part because of the members' prestige and political power. In his thorough and compelling study of the catastrophe, David McCullough concludes that "while there is no question that 'an act of God' (the storm of the night of May 30–31) brought on the disaster, there is also no question that it was, in the last analysis, mortal man who was truly to blame."[6] Wealthy mortal men did indeed evade any acknowledgment of real responsibility, but one hopes that the tragedy of Johnstown will never be forgotten and will be more frequently remembered with the trepidation it warrants.

An explanation for the title of Cambor's novel, *In Sunlight, in a Beautiful Garden*, is implied in the work's epigraph, which is taken from Claude Debussy's opera *Pelléas et Mélisande* (1902), with libretto by Maurice Maeterlinck: "I have been watching you: you were there, unconcerned perhaps, but with the strange distraught air of someone forever expecting a great misfortune, in sunlight, in a beautiful garden." The epigraph perfectly captures the sense of dread that pervades the entire novel.

The narrative of *In Sunlight, in a Beautiful Garden* largely centers on the members of two socially very different families. The Fallon family—Frank, Julia, Caroline, and Daniel (two other children perished in a diphtheria epidemic in 1879)—live in Johnstown and have achieved a comfortable middle-class existence through Frank's labor at the Cambria Iron Company. The Talbots—James, Evelyn, and Nora—are members of the South Fork Fishing and Hunting Club. James is a relatively successful lawyer, who is unhappily but cravenly drawn into the process of corruption that surrounds the legal approval of the club's siting and construction and the mistakes made with both the dam and the reservoir.

The novel begins on Memorial Day 1889 but quickly turns to the previous history of the dam and the club. There are frequent allusions to acts of corruption and cost cutting—major and minor—and the narrator eventually concludes about the wealthy club members that "these men who kept obsessive track of all their holdings, their vast properties, had no interest in the safety or the structure of the dam. It was as if no people lived below it, no world existed in the mountains but the one they were creating. Someone should have been assiduous, careful. Someone should have been watching."[7] But as became clear, no one was, as is often still the case in similar situations today. Studies by historians like McCullough bear out the narrator's implicit condemnation of the club's privileged members.

Despite their very different social backgrounds, Nora and Daniel—two unusual, sensitive, and intelligent young people—eventually get to know one another, a process that takes years. In the meantime, Daniel has become more and more concerned with the deaths that have occurred at the iron mill through negligence, and Nora's father, James, has developed increasing worries about the safety of the dam, worries he confesses to Nora. His concerns become increasingly obsessive, with good reason, as it turns out.

When the dam breaks, the water takes an incredible toll: "It combined with everything that lay in its path, hurling into the narrow valley, churning, flinging, and scouring a high wide path, now narrowing, now broadening, depending on the landscape it encountered, growing black, huge, monstrous, and fetid with debris. A black mist, a 'death mist,' as it was later called, hovered over it as it collected barbed wire, locomotives, railroad tracks, pulverized frame houses, keys and hobbyhorses, window glass, factory boilers, fuel."[8] Deaths abound; one of every nine residents perishes. Daniel instinctively attacks Nora as a representative of the guilty elite when he finds her at the club, and Nora does not resist: "He struck her, and she did not raise her hand to stop him."[9]

Daniel and Nora never see each other again. James eventually commits suicide, after trying unsuccessfully to get at least a few club members to share his sense of guilt and acknowledge the deliberate cost-cutting construction errors made in reconstructing the dam. But the club members maintain a careful silence and destroy, compromise, or hide evidence that might suggest any responsibility on their part. A postlude to *In Sunlight, in a Beautiful Garden* in 1917 ends with Nora's son, Samuel, leaving to fight in World War I. The sense of impending doom and despair with which the novel begins remains at the end, albeit for very different reasons. The prospect of imminent and unnecessary death haunts the novel from beginning to end.

Coopey's *Waterproof* is a tale of revenge and resilience. Alternating between 1891 and 1939, the novel provides a retrospective account by Pamela Gwynedd McRae, a survivor of the flood. Two years after the flood, Pamela's mother exists in a virtually catatonic state, Pamela's younger brother has drowned, and her father has disappeared, as has her former fiancé, Davy Hughes; but Pamela still struggles to create a meaningful life for herself. Davy, Pamela eventually learns, is now involved with a group that includes her father and is plotting revenge against the members of the South Fork Fishing and Hunting Club. The group engages in acts of petty sabotage—bombings, burning buildings, doing minor harm to a child of a club member. Eventually, Davy is exposed and arrested. He explains his motives to Pamela: "I'm so full of hate I can't think about anything else . . . I want to make them pay. Make them remember Johnstown every day of their rotten, rich lives."[10]

Pamela's attitude is very different. Although she does not gloss over or forget the horrors of the flood, instead at times providing painful and horrible descriptions of the events, Pamela is a resourceful young woman who eventually takes up a journalistic career and later marries an equally decent and upright fellow journalist. Pamela refuses to let the flood determine the course and tenor of her existence and of the many years that remain to her. She implicitly explains the novel's title, *Waterproof,* through a metaphorical comparison of herself with Davy, who possesses a very different attitude: "Right now the flood owned both of us, but I couldn't let that be the end of it. I had to struggle against it—make myself waterproof. And Davy, too, if only he'd let me."[11] But he won't. Pamela tells Davy: "We don't have to let the flood own us. We could beat it if we really tried."[12] And Pamela not only tries but succeeds. A strong and moral person, she goes on to live a satisfying life of rectitude and personal and professional success. Davy, in contrast, ironically becomes a cynical and wealthy man with principles as compromised as those of the men against whom he plotted revenge.

Aswan Adventures

The failure of the South Fork Fishing and Hunting Club did actually take place, and a century and a half later the dramatic catastrophe that ensued still provides material for dire and disturbing fiction. Other dams written about just as often have not failed but still generate continual fears of failure or attack, despite their apparent resilience. Among these, the best known is Egypt's Aswan High Dam.

The Aswan High Dam, the largest embankment dam in the world, was completed in 1970. Its completion was preceded by decades of strife, frustration, and fear. Disputes about the Suez Canal, which had originally been built between 1859 and 1869 to provide a human-made route from the Mediterranean to the Red Sea, were exacerbated by multinational Cold War political tensions in the 1950s. Frustrated by what he regarded as excessive American interference, Gamal Abdel Nasser declared the Suez Canal nationalized. This led to the Second Arab-Israeli War (the first had been in 1948). With support from the United Kingdom and France, Israel invaded Egypt in late 1956. Pressure brought to bear by the United States and the USSR, despite Cold War tensions between the two, brought the crisis to an end, at least temporarily.

Other regional crises followed—the Six-Day War in 1967 and the Yom Kippur War in 1973. A treaty brokered in the late 1970s led to an uneasy peace between Israel and Egypt that has, to the surprise of many, lasted in a somewhat unsteady manner. There have, however, still been a host of unpleasant, dangerous, and potentially threatening episodes, including many violent border incidents. In 2001 the hawkish Israeli foreign minister Avigdor Lieberman accused Egypt of plotting a surprise attack against Israel and suggested bombing the Aswan High Dam. In 2011 Egyptian anger about perceived Israeli malfeasance led to an attack by thousands on the Israeli Embassy in Egypt. Small wonder, then, that in this ongoing context of mutual lack of political confidence there are constant fears that some unfortunate day, the Israelis or someone else might destroy the Aswan Dam and unleash an unbelievable environmental catastrophe, millions of deaths, and massive political and social tribulations on Egypt and the world.[13]

The novels written about a possible attack on the Aswan Dam fall into a number of categories. There are multiple thwarted disaster narratives, in which indomitable Israelis combat evil and frequently sadistic Arab terrorists,

Dam Failures, Real, Imagined, and Ecotage-Inspired   153

who scheme to blow up the dam and cast responsibility on the Israelis. The Cold War haunts the background of many such novels. A good example is Andrew Sugar's *The Aswan Assignment* (1974), in which an Arab terrorist group known as Black February plans to destroy the dam and blame Israel for it, thus inspiring an end to Israel's existence because of world condemnation. The terrorists do not actually intend to blow up the dam, only to make it look as if there has been an Israeli attempt to do so by leaving two dead Israeli underwater demolition experts floating in the water near the dam. Fortunately, Israeli commando Dov Abrams parachutes into Egypt, rescues the kidnapped Israeli demolition experts, kills several terrorists, and engages in a bit of pleasurable sex with a female Israeli agent. All ends well, if somewhat predictably.

A similar plot drives Gar Wilson's *Aswan Hellbox* (1983), in which a terrorist group known as the Black Cobras plans to infiltrate Sudan, stage an attack across the Egyptian border, and destroy the Aswan High Dam—thus leading to war between Egypt and Sudan and general social and political chaos. An American group, Phoenix Force, thwarts this dastardly plan, behind the scenes of which lurk cunning Soviets determined to foment world havoc. The Cold War is in full view in *Aswan Hellbox*. Once again, though, the evil terrorists fail miserably.

Novels like *The Aswan Assignment* and *Aswan Hellbox* undoubtedly provide an entertaining, excitement-packed read for some. Other novels about Aswan have taken a less predictable and more interesting narrative tack. Michael Heim's *Aswan!* (Assuan Wenn der damm bricht, 1971) imagines a situation in which the dam itself is in real danger of failing; water is escaping belowground, and the dam is in desperate need of stabilization. This creates a geopolitical threat of terrorism and war. At the beginning of the novel, Star of David flags are found impaled in the dam crest, suggesting—falsely, it appears—Israeli responsibility for the problems that are emerging. Other suspicious incidents begin to evoke fear worldwide. Both Israeli and Egyptian authorities begin to consider bombing the dam in a potential effort to avert the death of millions by providing advance warning of the dam's destruction. A revealing reference is made to the fate of Germany's Edersee Dam in 1943.

There are multiple suspects in *Aswan!* Who might wish to blow up the dam. In addition to hawkish Israelis, there are displaced Nubians, unhappy peasants, and Islamic extremists. The Egyptian authorities themselves disagree about the dam's usefulness. A Colonel Shuker demands of President Sadat: "Why didn't you and Nasser build a Great Wall of Egypt along the Libyan Desert to keep out sandstorms and fleas? Your High Dam is a monstrosity—it's unnatural in the truest sense of the word."[14] General Shafai declares: "The dam must go. It must be demolished before it's swept away."[15] It is interesting that such criticisms of Aswan's viability and susceptibility to deterioration were already beginning to be expressed in the early 1970s, albeit in fiction by non-Egyptians. The great enthusiasm that greeted the dam's opening in the 1960s did not last. In this sense, the Aswan High Dam is a good example of thoughtlessness enhanced by the Cold War.

Other members of the Egyptian high command in *Aswan!* Offer majority support for attempting to discover a way to stabilize the Aswan High Dam. In the novel, not all Egyptians share the views of Shafai and other military commanders. Multiple techniques are considered, and the aid of a wide range of European and American experts is enlisted. Meanwhile, in Israel, public opinion swings dramatically in the direction of helping the Egyptians. Elaborate operations are planned, and Israel assumes a highly positive role. This does not, however, prevent some fearful Egyptians from launching pogroms against Israelis in various Egyptian cities.

Meanwhile, a peculiar film (a kind of artificial growth) has developed that is inhibiting evaporation from the reservoir behind the Awsan High Dam. Mario Angelo, the Italian scientist who could solve the problem of the film's origin, commits suicide, mocked by fellow scientists who doubt the efficacy of the evaporation inhibition technique he has devised for reasons completely unrelated to the safety of the Aswan High Dam but that could salvage the situation. Eventually, the dam collapses and catastrophe follows. *Aswan!* Ends on an understated but dire note: "Dr. Mario Angelo was buried at San Michele that morning. The gravediggers had prepared Plot No. 16, Row 13, Section Q, to receive his coffin. Twelve hours later, the water level in Venice rose six inches."[16] Six inches. Who knows how much of the world is now doomed, but widespread catastrophe is certain.

Two other novels about threats to the Aswan High Dam that offer an equally interesting look at the potential complications surrounding the dam are John Rowe's *The Aswan Solution* (1979) and R. E. Harrington and James A. Young's *Aswan High* (1983). The major protagonists of *The Aswan Solution* are David Laker, a Jewish electronics expert who works for the Rand Corporation

Dam Failures, Real, Imagined, and Ecotage-Inspired 155

on a device capable of jamming antiaircraft missiles, and Miriam Heller, an Israeli agent. The Israelis plot to seize the Aswan High Dam, place a nuclear bomb in it, and hold the Egyptian military hostage by threatening to detonate the bomb and drown Egypt. Their major goal is to prevent nuclear war and Israel's destruction while knocking out Egyptian nuclear capabilities. Laker's help is essential to this complex and convoluted plot.

As planned, Miriam successfully evokes Laker's romantic interest, and the Israelis stage a purported Arab terrorist ambush to ensure the American's willingness to work with Israeli intelligence. The dam is seized and an international cast of characters gets involved in the increasingly threatening developments. Everything seems to be going well according to the plan for saving the dam, but at the last moment things go terribly wrong and the dam is destroyed. Now what? Will Israel be condemned like the Nazis or admired like the Americans for exterminating the Japanese at Hiroshima and Nagasaki, wonders the narrator of *The Aswan Solution*. The novel ends on a dismal note. The Aswan solution of the novel's title has been revealed as no solution but as a catastrophe.

Harrington and Young's *Aswan High* also examines the potential for a disastrous outcome for world stability. In the novel's prologue, set in 1967, three strangers pay an Egyptian power shovel worker to bury two supposed rocks inside the Aswan High Dam. Years later, in 1984, an American archaeologist, Phillip Burton, becomes involved with Rebekah Altman, daughter of the scientist and scholar Isaac. Isaac, who is eventually revealed as one of the three strangers, dies shortly after Phillip's arrival in Israel, unable in his final moments to explicitly communicate a terrible and important bit of information. His enigmatic message states simply: "To R [Rebekah] and P [Phillip] See Moses and Ben *Danger* Ahmend Aswan destroy code."[17] Isaac is desperately opposed to the collapse of the Aswan High Dam.

Schneor Ben-Solomon and Moses Sacher, longtime colleagues of Isaac and the other two of the three strangers present in 1967, clearly know something of great importance but refuse to reveal it. Eventually, though, after the Arab terrorists intrude on the scene, the secret is finally exposed. Back in 1967, the three Israelis had arranged for the placement of a bomb in the Aswan High Dam, but each knows only a part of the code that will activate its detonators. In familiar fashion, the Arab terrorist Shukri Rifaat now wants to blow up the dam, cast the blame on Israel, and thereby cause the destruction of the

156   DAM FAILURES, REAL, IMAGINED, AND ECOTAGE-INSPIRED

country; the likely collateral death by drowning of millions of Egyptians is of little consequence to this rabid fanatic.

Unfortunately, Rifaat does discover the rest of the code, and a hideous explosion takes place. But things do not turn out as badly as they might have. International aid efforts are mobilized, the role of the terrorists is exposed, and Rifaat is killed. All goes about as well as it could: "The disaster relief provided to Egypt beginning in the autumn of 1984 was the largest co-operative international effort ever mounted."[18] The countries of the Middle East band together; North Americans, British, Russians, and Israelis work side by side to salvage the situation. Rebekah and Phillip's romance has a promising future. In contrast to the catastrophe that marks the end of *The Aswan Solution*, *Aswan High* offers compelling reasons for hope for a brighter future for the world, for the potential for meaningful international cooperation unmarred by internecine hatreds.

Ataturk and Disaster Managed and Mismanaged

The Aswan High Dam may be the best-known potential source of international friction among dams in the Middle East, but it is by no means the only one that causes concern. The Ataturk Dam, located in southeastern Turkey in Anatolia, is also a cause for worry. Completed in 1990, the Ataturk Dam was originally called the Karababa Dam but was renamed in honor of Mustafa Kemal Atatürk, the founder of the modern Turkish Republic. The disputes related to the Ataturk Dam involve primarily Syria and Iraq and the fact that the dam reduces the flow of the Euphrates to the latter in an already heavily water-needy area.

Two of the novels that address the potential conflict associated with the Ataturk Dam are Jeff Rovin's *Acts of War* (1997) and Richard Sherbaniuk's *The Fifth Horseman: A Novel of Biological Disaster* (2001). *Acts of War* is one of a series of novels called Op-Center created by Tom Clancy (of terrorist novel fame) and Steve Pieczenik. *Acts of War* focuses on a Syrian Kurdish militant group that begins its activities by stealing a Turkish helicopter and then bombing the Ataturk Dam. The group's ultimate, and familiar, goal is to provoke war in the Middle East. Op-Center is a sophisticated surveillance and crisis management operation implicitly driven by Cold War tensions and Middle Eastern antipathies. Although the terrorists seize hostages, in the end, of course, the savvy Op-Center personnel triumph.

The Fifth Horseman involves a much more elaborate biochemical plot. Dr. Mike Zammit of the International Environmental Response Team (INERT) discovers a plan devised by renegade Russians and other miscreants—including a Russian mole at work within INERT—that is intended to poison water in the Ataturk Reservoir and cause thousands of deaths, thereby provoking a regional war. Once again, the Cold War and Middle Eastern disputes lurk in the background. The details involve introducing into the reservoir a mutated cyanobacteria that will produce immense amounts of toxins. A growing gelatinous mat of poisonous algae will facilitate this diabolical project, the brainchild of, among others, a sadistic Russian saboteur, Ruslan Glinka, and an equally pathological wealthy Omani, Abdul Jamal. Syrian and Iraqi participation complicates the picture even more, as does the involvement of a renegade Russian general, Vladimir Bled, who seeks revenge for what he regards as humiliation of the former Soviet Union by the West. And then there is the evil Russian Orthodox monk Smegyev, who is working with Bled to bring the plot to fruition.

The science brought to bear to stop the cyanobacteria from taking over the reservoir involves a complex method of starving the bacteria. An antidote is discovered that binds to the chlorophyll on the cell periphery and makes sunlight poisonous to the cyanobacteria. Time is of the essence, though, for the bacterial spores must be destroyed before they grow wings and disperse across the landscape. As is so often the case in Cold War–related narratives, luck is on the side of the good guys (that is, the West), and for the moment all is well. Pure water and hydroelectric power now flow from the Ataturk Dam, and Abdul Jamal dies a fittingly horrible death. The Iraqis and Syrians remain within their own borders, held in check by UN sanctions and a huge North American Treaty Organization (NATO) force in the area. Once again, an evil terrorist plot has been thwarted and general regional peace established. Implicitly, the West has inflicted a defeat on mad Soviet cold warriors and their Middle Eastern collaborators.

Yangtze Horrors

The Three Gorges Dam, which excited much controversy before and while it was being built, is also a frequent subject for catastrophic, but thus far curiously unsatisfying, narratives. Peter Tonkin's *Red River* (2010) begins promisingly but ends weakly. Unlike many of the novels discussed above, *Red River* features a natural disaster involving the Three Gorges Dam rather than an act of terror. Scientists Robin Mariner and her husband, Richard, are on a boat at the mouth of the Yangtze when the Three Gorges Dam collapses because of an immense earthquake. Mafiosi-like Triad gangs (a transnational Chinese organized crime syndicate) and pirates are heavily involved in the ensuing adventures, in part because of the recent discovery of a hugely valuable statue of Genghis Khan. The novel's major plot involves a race against time to restore an elaborate pumping system that will save Shanghai from being drowned. The novel ends with a refreshing sigh of narrative relief when the city is saved, but it is disappointing in its lack of attention to what one knows must be catastrophic damage up and down the Yangtze and any clear details as to its resolution. All is well, supposedly, but one wishes for more narrative information about the consequences and ways with which they are dealt.

Thomas V. Harris's *Three Gorges Dam* (2018) links Uighur terrorism and geological failings. An earthquake leaves the Three Gorges Dam still standing, but aftershocks cause serious damage and deaths. A major message of the novel is that a dam should never have been built at such a geologically vulnerable site. Finally, overtopping causes the dam to collapse. The good guys are rescued by helicopter, but a massive tsunami destroys vast numbers of people, vehicles, and buildings. The future is uncertain and details are unexplored, just as in *Red River*, but the element of romance makes the uncertainty slightly more palatable.

Both *Red River* and *Three Gorges Dam* leave the reader with a feeling of frustration because the supposedly happy endings leave so much detail to the imagination. What happens to the Yangtze after the collapse of the Three Gorges Dam? Are there massive international salvation efforts? Is the final message that romance is always most important and environmental catastrophe incidental? The reader of both *Red River* and *Three Gorges Dam* is left with many questions, not the least of which is how seriously one should worry about a realistically dire fate for the Three Gorges Dam.

Environmentalism and Its Discontents: *The Monkey Wrench Gang*

Conservation. The word itself evokes inspiring images—of protection, of salvation, of prohibition of waste, of the preservation of all that is beautiful.

Unfortunately, for humans, the notion of sustainable use of nature is a slippery slope, a twisted path down which human beings have been sliding for millennia. An often-cited verse from the Bible that was mentioned earlier comes from Genesis 1:28–30: "God blessed them and said to them, 'Be fruitful and increase, fill the earth and subdue it, rule over the fish in the sea, the birds of heaven, and every living thing that moves upon the earth.' God also said, 'I give you all plants that bear seed everywhere on earth, and every tree bearing fruit which bears seed: they shall be yours for food. All green plants I give for food to the wild animals, to all the birds of heaven, and to all the reptiles on earth, every living creature.'" This passage raises several questions about conservation—conservation of what, for whom, and why. What exactly did God mean? Are human beings truly and rightfully in charge of nature?

In later centuries, as we saw in chapter 1, many interpreted God's exhortation as one of divinely ordained human stewardship of the land and its sensible use, with the emphasis on use—often verging on exploitation—and the meaning of sensible a matter of interpretation. The naturalist John Muir (1838–1914) was one of the first to adopt a different stance and advocate aggressively for preservation of the North American wilderness. One of his most heartfelt causes was the Hetch Hetchy Valley in what is now northwestern Yosemite National Park. There were those who wanted to develop Hetch Hetchy for irrigation and the provision of water for San Francisco. Despite Muir's efforts, in conjunction with the Sierra Club, for more than a decade to prevent the construction of a dam, a law authorizing the Hetch Hetchy Dam was passed in 1913; the dam actually exists.[19]

A prominent admirer of Muir, David Ross Brower (1912–2000), executive director of the Sierra Club from 1952 to 1969, led an analogous campaign against building a dam at Echo Park and Dinosaur National Monument on the Colorado River.[20] Brower and his fellow activists viewed this project, among other plans for dams on the Colorado, as an assault on the purity and beauty of the Grand Canyon. In 1956 their efforts succeeded, and the project was abandoned. Not all their attempts to halt the building of dams were successful, though. For years Brower and his fellow advocates fought relentlessly against the completion of Glen Canyon Dam, an immense concrete gravity dam on the Colorado River in Arizona about sixty miles north of the Grand Canyon that Brower considered perhaps the worst environmental mistake the US government had ever made (figure 5.1). Glen Canyon Dam

FIGURE 5.1. André Payan-Passeron, *Glen Canyon Dam*, photograph, December 1, 2012. Wikimedia Commons, https://commons.wikimedia.org/wiki/File:Le_barrage_de_%22Glen_Canyon_Dam%22_sur_le_Colorado_(USA).JPG.

was built between 1956 and 1966, despite widespread and very vocal opposition. Behind it lies Lake Powell, one of the largest human-made reservoirs in the United States. Glen Canyon Dam remains a source of controversy to this day.[21]

Brower's assessment of the merits of Glen Canyon Dam was comparatively mild in comparison to that of the feisty environmentalist Edward Abbey (1927–89). Abbey, who had fallen deeply in love with the area as a teenager and later worked briefly as a park ranger at Arches National Park in southern Utah, took a trip with a friend down the Colorado before the construction of Glen Canyon Dam. His essay on the trip, published in 1968, begins with a vehement diatribe:

> The beavers had to go and build another goddamned dam on the Colorado. Not satisfied with the enormous silt trap and evaporation tank called Lake Mead (back of Hoover Dam), they have created another, even more

Dam Failures, Real, Imagined, and Ecotage-Inspired 161

destructive, in Glen Canyon. This reservoir of stagnant water will not irrigate a single square foot of land or supply water for a single village: its only justification is the generation of cash through electricity for the subsidy of various real estate speculators, cottongrowers and sugarbeet magnates in Arizona, Utah and Colorado; also, of course, to keep the engineers and managers of the Reclamation Bureau off the streets and out of trouble.

The impounded waters form an artificial lake named Powell, supposedly to honor but actually to dishonor the memory, spirit and vision of Major John Wesley Powell, the first American to make a systematic exploration of the Colorado River and its environs. Where he and his brave men once lined the rapids and glided through silent canyons two thousand feet deep the motorboats now smoke and whine, scumming the water with cigarette butts, beer cans and oil, dragging the water skiers on their endless rounds, clockwise.[22]

This is an ugly picture of thoughtless human misuse of the Colorado for selfish pleasures and pollution.

In "Down the River," Abbey also fantasizes in detail about the potential failure and destruction of Glen Canyon Dam, possibly by outright ecoterrorism:

The dam will never be completed, they'll run out of cement or slide rules, the engineers will all be shipped to the Upper Volta. Or if these fail some unknown hero with a rucksack full of dynamite strapped to his back will descend into the bowels of the dam; there he will hide his high explosives where they'll do the most good, attaching blasting caps to the lot and with angelic ingenuity link the caps to the official dam wiring system in such a way that when the time comes for the grand opening ceremony . . . the button which the President pushes will ignite the loveliest explosion every seen by man, reducing the great dam to a heap of rubble in the path of the river.[23]

Talk may be cheap, but Abbey was not just talking. Angered in part by the depredations of the US military-industrial complex in the mid- to late twentieth century and by Vietnam, he was driven to start a sustained environmental movement to counteract what he regarded as US governmental failures. Abbey's efforts and those of many others over the past few decades have led to the creation of a network of would-be protectors of the environment. What to call such people? As in many other instances, the precise terms used may suggest approval, bias, or condemnation. In *Eco-Warriors, Nihilistic Terrorists, and the Environment* (2011), Lawrence E. Likar distinguishes

162 DAM FAILURES, REAL, IMAGINED, AND ECOTAGE-INSPIRED

between ecoterrorism and environmental terrorism. Of the former he says, "*Ecoterrorism* has been defined by the FBI as 'the use or threatened use of violence of a criminal nature against innocent victims or property by an environmentally oriented subnational group, for an environmental-political reason.'"[24] For Likar, *environmental terrorism*, in contrast, can refer to either "an attack against persons or property using an element of the natural environment (e.g. water or fire) as a weapon or an attack directly against a natural resource (e.g., forest or ocean reef) for the purpose of instilling fear in a human target."[25] This appears to be a very fine distinction. Likar wisely notes that terrorism is always a difficult term to define and that self-declared fighters for the environment may refer to themselves as eco-warriors, which has a more positive connotation than terrorist.[26] He also points out that such individuals may operate as "lone wolves," outside the context of an organized group.[27] This observation has sinister implications and points to the possibility of machinations by individual maniacs.

Enter the amazingly influential Abbey, who favored violence against property but not against people.[28] The term Abbey made famous is *monkey-wrenching*, which comes from the title of his best-known novel, *The Monkey Wrench Gang* (1975). James M. Calahan, the author of a biography of Abbey, declares that the concept of monkey-wrenching, according to Abbey, refers to attempts by small, close-knit groups to sabotage invidious machinery without hurting people.[29] Likar defines monkey-wrenching as the destruction or sabotage of equipment or facilities that harm the environment.[30] Like Calahan, Likar argues that Abbey favored destruction of property but not of people.[31] Rik Scarce, author of *Eco-warriors: Understanding the Radical Environmental Movement* (2006 [1990]), employs a slightly different, and less kind, definition. In general usage, the term *Luddite* refers disparagingly to someone opposed in knee-jerk fashion to new technology or ways of working. Scarce writes: "'Luddite' has . . . come to mean a technophobe or saboteur. Ecological sabotage, or 'ecotage,' is the name environmental Luddites give to destruction of technology in defense of nature."[32] Clearly, Scarce does not approve of such unthinking violence.

The four major protagonists of *The Monkey Wrench Gang* are "Seldom Seen" Smith, a wayward Mormon river guide; Doc Sarvis, a wealthy elderly surgeon with a strong ecological commitment; Bonnie Abbzug, a transplanted Brooklynite and Sarvis's assistant and girlfriend; and George Hayduke, a

Dam Failures, Real, Imagined, and Ecotage-Inspired   163

Vietnam veteran and Green Beret who shares the same strong distaste the others do for environmental despoilment by machines. Hayduke's experience in Vietnam is significant and points to the disappointment in American ways and Cold War machinations that some Vietnam veterans and others came to feel. Together, the group engages in acts of ecotage against billboards, road-building equipment—especially bulldozers—survey markers, and bridges as preparatory punitive raids for action to be taken against the greatest objects of their loathing: the Glen Canyon Bridge and Dam. Smith calls Lake Powell "the blue death," in a reference to the reservoir's "dead water."[33] We have seen such connotations of death associated with reservoirs before. The foursome's most magnificent supposed achievement is the imaginary destruction of the Glen Canyon Bridge:

> Suddenly the center of the bridge rose up, as if punched from beneath, and broke in two along a jagged zigzag line. Through this absurd fissure, crooked as lightning, a sheet of red flame streamed skyward, followed at once by the sound of a great cough, a thunderous shuddering high-explosive cough that shook the monolithic sandstone of the canyon walls. The bridge parted like a flower, its separate divisions no longer joined by any physical bond. Fragments and sections began to fold, sag, sink and fall, relaxing into the abyss. Loose objects—gilded scissors, a monkey wrench, a couple of empty Cadillacs—slid down the appalling gradient of the depressed roadway and launched themselves, turning slowly, into space. They took a long time going down and when they finally smashed on the rock and river far below, the sound of the impact, arriving much later, was barely heard by even the most attentive.[34]

The cultural and political influence of *The Monkey Wrench Gang* was and remains immense. In the immediate aftermath of the publication of the novel, Abbey's "writing spawned the formation of the radical environmental group Earth First!, which found itself the target of FBI infiltration in the 1980s and a federal conspiracy trial in Prescott, Arizona, in the '90s."[35] Abbey himself was also under observation for decades by the FBI; in addition, as alluded to above, American disputes over Vietnam contributed to the tension surrounding the values or lack thereof of US technology.

In the novel, three of the participants in the Monkey Wrench Gang are tracked down by the police, sent to court, and convicted, but only at a relatively

164 DAM FAILURES, REAL, IMAGINED, AND ECOTAGE-INSPIRED

minor level. Because of both a lack of hard evidence and skillful wheeling and dealing by Doc Sarvis, he, Abbzug, and Smith are sentenced to concurrent terms in the Utah State Prison of not less than one year and not more than five years. The judge suspends the sentences because of the defendants' records and other circumstances but orders them to spend six months in jail and then serve four-and-a-half years each on probation. Two years later Hayduke, whom everyone has presumed dead, shot by the authorities, reappears. He has a new identity and new job as a night watchman and is still a convinced monkey wrencher. Ironically, in the final pages of the novel, it is revealed that it was not, in fact, the Monkey Wrench foursome who blew up the Glen Canyon Bridge. Who did is unknown, but the fact that someone else did suggests that ecotage is becoming a movement, not just the crazy goal of a small group of misfits.

On the surface, ecotage, to use the most neutral official-sounding syn-onym for monkey-wrenching, sounds relatively benign—a kind of well-meant vandalism (contradictory as that might sound) with no intention of harming human beings. Unfortunately, theory and practice, as is well-known, often diverge. Novels written later that were not wholly unsympathetic to Abbey's immensely influential point of view about protecting nature from human ruin (novels can inspire political action) demonstrate that ecotage can end badly, with not only dangerous destruction but multiple deaths. Magnificently effective in demonstrating this point is Gary Hansen's *Wet Desert: Tracking down a Terrorist on the Colorado River* (2007).

Wet Desert: Dreams Gone Wrong

Gary Hansen is an electrical engineer and lover of the West, especially the Colorado River. His professional expertise enabled him in *Wet Desert*, his first novel, to explore the intricacies of what actually might happen to the Colorado and its environs were a large dam to be destroyed. The story is both informative and gripping,

The hero of *Wet Desert*, and he is a hero in the true sense of the word, is Grant Stevens, a mid-level manager at the United States Bureau of Reclamation. Grant's passion is the actual construction of dams, an enter-prise now mostly relegated to the North American past because of both fewer desirable dam sites and environmental objections. As his mentor, who had helped design the Glen Canyon Dam, tells him: "Face it, Grant, there ain't gonna be no more dams in America. It's over. It's not considered environmentally correct to build dams anymore."[36] Grant's only recourse is to protect already constructed dams.

Through a peculiar set of circumstances, Grant has to take charge in the immediate aftermath of the destruction of the Glen Canyon Dam by a terrorist. In keeping with one of Likar's hypotheses, the terrorist in question is a "lone wolf" who remains anonymous and personally somewhat peculiar throughout the novel. He is fanatically obsessed with the Colorado River, making a living as a casino technician but spending all his free time alone in the desert. He imagines the pre-dam beauty of Glen Canyon in all its glory:

> What he had never seen, unfortunately, were the canyons themselves, under all the water. Only about a thousand people ever had, before the dam buried them forever. He read accounts of people lucky enough to have explored them including John Wesley Powell himself. They declared Glen Canyon one of the most beautiful places on earth. They described pink undulating sandstone walls, some striped, with rain forest–like jungles in some of the side canyons, and green fractures high on the walls nourished by seeping springs. The endless carved rock canyons contained lush overhangs and rock amphitheaters. But now it was all gone, forever. It made his stomach tighten every time he thought about it.[37]

The terrorist realizes that because of vested financial interests, decommissioning Glen Canyon Dam is nothing but an idealized fantasy, and so he comes up with a different and more dramatic and dangerous plan—namely, to blow up the dam.

Many of the novel's characters are tourists enjoying the pleasures of boating on Lake Powell and other Colorado River reservoirs. The protagonist of *Wet Desert* is not so naive as to anticipate a complete absence of deaths. In fact, his biggest regret is that many of his potential victims are likely to be environmentalists and nature lovers like himself, not the bureaucrats he loathes: "He would feel much less guilt about killing them [the bureaucrats]. A scene played in his mind where bureaucrats on deck raised their arms to fend off the water and their ultimate death. That scene felt good. Unfortunately, the politicians were in Washington, not on a yacht in the river. It was his people down by the river."[38] But that is how it has to be, he thinks, and collateral damage that results in many deaths is unavoidable.

In keeping with Abbey's priorities, the terrorist's first object of attack is the Glen Canyon Dam. With his technical background, he recognizes that destroying a dam is not as easy as it might seem. After much thought, extensive research, and thorough planning, he comes up with a workable scheme. Throughout *Wet Desert*, much detail is provided about both the terrorist's destructive activities and Grant's skillful management of ways to counter the problems a cascading series of dam failures creates. In *Wet Desert* Hansen captivates the reader with the details, maintaining great suspense. He also humanizes the narrative by discussing the adventures of several terrified groups of characters caught in the flooding that begins when the dam is destroyed. These stories of individuals, some of whom perish, make *Wet River* vivid from a human standpoint as well as technologically fascinating.

Glen Canyon is just the beginning for the terrorist. His ultimate plan, a scheme of massive proportions, involves destroying dams all the way down to Mexico and restoring the river landscape to its original condition. But Grant is a more than worthy opponent with remarkable expertise. He is able to work well with other dam employees and overcome the thoughtless objections of ignorant bureaucrats from the Bureau of Reclamation, state governments, and the FBI. The major challenge facing Grant is a race to make arrangements at dams downstream from Glen Canyon that will prevent further destruction, flooding, and deaths. The water released from Lake Powell must somehow be stymied, the catastrophe lessened.

Grant is in command of facts and figures and has brilliant assistants who can do relevant and important calculations with amazing speed. He is also assisted by many local volunteers. Much of the work to be done involves blowing up dam spillways to allow the water to pass down the river in a more measured way, to prevent sudden overtopping and rapid destruction of more dams. Recall the horror of the Johnstown Flood. A key goal is also to raise temporarily the height of the Hoover Dam to prevent overtopping. This also calls to mind the reasons for the failure of the South Fork Dam. The unusual and initially resisted solution is to create a dike on top of the Hoover Dam made of sandbags. Another tactic, one unpopular with many dam engineers in the novel, is to deliberately blow up smaller dams downstream to slow the river's movement. All of these tactics work, however, and minor losses help avoid major catastrophes.

In the meantime, the FBI is frantically investigating to find out if there is a group or a lone terrorist at work and to identify them. When the terrorist blows up the California Aqueduct and attempts to blow up the All American Canal, it is Grant who finally grasps the terrorist's ultimate plan, which is to restore the badly eroded Mexican delta at the base of the Colorado River. Ironically, once Grant understands this, his sympathy for the terrorist increases: "Grant felt funny. They had just cracked the case wide open. The Colorado River bomber was an environmentalist. He was sure of it. Now the FBI would know where to look. They could track him down. But Grant didn't feel as good as he should, and he sensed that the others didn't either. It had been easier when they thought the bomber wanted to destroy, maim, or kill. Now the motive turned out to be restoring a wildlife habitat. Now what? Grant knew what they had to do, but his feelings had changed."[39] Grant understands that the terrorist will want to be at the exact spot where the former Colorado delta is restored and to be there when it happens. He and his companions fly into Mexico illegally, and Grant discovers and attempts to speak with the terrorist. But the terrorist deliberately allows himself to drown.

The epilogue to *Wet Desert* is as satisfying as it can be under the circumstances. Grant is put in charge of rebuilding the Glen Canyon Dam, but the federal government also makes an important compromise by recognizing the environmental catastrophe that has been wrought over the years on the Colorado and its delta. The plan is to restore the delta by allowing more of the Colorado River water to flow freely into Mexico than has been true in the past. It is not a perfect solution, but it is a significant step in the direction of improving the Colorado River environment.

Avenging the Columbia Salmon

In the dedication to *K Falls* (2001), Skye Kathleen Moody thanks, among others, "the late Edward Abbey, whose novel *The Monkey Wrench Gang* got it right."[40] The Ojibway prayer with which she prefaces the novel offers hope, ending with the plea "Grandfather/Sacred one, / Teach us love, compassion, honor / That we may heal the earth / And heal each other."[41] Throughout the novel, Moody's sympathies are clearly with the Native Americans and the salmon. The opening of the novel doesn't accord with these sentiments,

168 DAM FAILURES, REAL, IMAGINED, AND ECOTAGE-INSPIRED

however. The narrator ends a beautiful description of the river with a bleak statement: "The salmon's silver skin reflected so brightly that it blinded the farmers, who cursed the damage to their fields. After 'civilization' discovered the Columbia River, change came rapidly at the hands of people who had forgotten that Nature has a soul."[42] Particularly in the novel's later chapters, Moody includes an occasional lengthy epigraph that points to a picture of the Columbia River devoid of hope for survival of the salmon and for any actions of human goodness.

K Falls tells a story of a bizarre cast of characters involved in one way or another with salmon on the Columbia River. The K in K Falls is an abbreviation for Kettle Falls, one of the main Columbia falls fishing sites drowned by the construction of large dams like Bonneville and Grand Coulee in the 1930s through the 1950s. Kettle Falls now lies under Lake Roosevelt behind the Grand Coulee Dam. The Spokane and Consolidated Colville Indians suffered most from the loss of the falls.

The positive protagonists of *K Falls* include Venus Diamond and Louis Song, rogue undercover agents for the United States Fish and Wildlife Service who are in a race to discover who is responsible for recent failed attempts to blow up dams on the Columbia. They eventually work directly, but incognito, with the perpetrator to foil his evil intentions. The other protagonists in *K Falls* are a very mixed lot. There is the Dambusters Fund, which especially wants to breach Grand Coulee Dam so the salmon can once again reach their spawning grounds. Its program is essentially positive. The fund is concerned about the supersaturation of nitrogen that takes place when excess water flows over the spillways, as well as the fact that some fish are sucked into the dams' turbines. According to the narrator, there is a range of solutions to these problems: "Industry's solution: Spread nets across turbines. Catch young fish, and truck or barge them around the dams. Envirosolution: Remove dams, restore rivers, return salmon. Ecoterrorist's solution: Ka-boom."[43]

The Dambusters Fund is led by Fritz Fowler, who is assisted by, among others, Sue Ann Denny, a former wild and violent activist. Sue Ann, it turns out, is the adoptive mother of Darla, a naive but very intelligent young woman. Darla becomes involved with Gerald, a true ecoterrorist who, the reader eventually learns, has worked for the power company for several years and is thus very familiar with the ins and outs of the dams' maintenance system. Gerald is also an anarchist and a sadist, a "lone wolf" who bullies his female accomplices and forces them to demonstrate their commitment to the cause by walking barefoot across burning coals. Gerald supposedly cares about saving the river, but it is difficult to tell, given his pathological sadistic inclinations and penchant for intimidation.

Darla, who works for a bank, eventually suspects that her supposed mother has sent Gerald to seduce her because she works at a bank and can provide information to help with plans to fund the plot to blow up Grand Coulee. Darla, who is by no means an ecoterrorist in her inclinations, sends anonymous warnings to Venus about Gerald's plans to blow up the dam, including very specific details her intelligence helps her discover and clarify. The novel ends in a blaze of violence. Perhaps it is needless to say that the dire plot is thwarted and Gerald is killed. Ecoterrorism is stymied this time, but the novel does not end on the same positive note as does *Wet Desert*.

A River Out of Eden: Ecotage Can Be Gratifying

The final ecotage-related novel I wish to discuss, and in some ways the most curious, is John Hockenberry's *A River Out of Eden* (2001). The novel has a strange cast of characters. There is Francine Smohalla, who works at a United States Army Corps of Engineers salmon hatchery next to the Bonneville Dam complex. A marine biologist, Francine is half Chinook (Smohalla was a late nineteenth-century prophet associated with the Native American Dreamers movement in the Pacific Northwest, which sought to restore Native ways and escape white influence). Her father, Charley Shen-oh-way, disappeared when she was a baby and is presumed to be dead. Her mother, Mary Hale, daughter of a deceased businessman who made his wealth through heavy involvement with dam building on the Columbia, has been virtually comatose for over thirty years. Francine lives with her mother in a mansion above the Columbia. Francine understands that the fish ladders at Bonneville do not work, but she nonetheless tries desperately to help the hatchery fish, mainly by overseeing their trucking transport past danger points on the river.

Duke McCurdy, son of a survivalist and member of the Aryan Nation, is a boating lover whose mother was killed by federal agents when he was a young boy. Duke devotes himself to his father and spews racist tripe over the family-owned radio station in the persona of Tommy Liberty. Duke is not a convinced Aryan, but he is tied to his father because of the loss of his mother. Then there is Jack Charnock, a disillusioned nuclear device engineer and designer who is about to retire from the Hanford Nuclear Reservation on the Columbia River, which played a crucial role during World War II and is now heavily contaminated with radioactive materials. Jack has worked secretly for years on miniaturizing a plutonium implosion device. His bitterness stems largely from the heavy exposure to radioactivity his father suffered at Hanford and his own son's accidental death, also at Hanford. His final object is destruction of the Columbia dam system.

Francine and Duke meet because they both find mutilated dead bodies—Francine at Bonneville, Duke at the opening of the Columbia River. The bodies, which are soon joined by others, are those of hatchery fish truck drivers. The mutilator, it emerges, is Francine's long vanished father, who is now in the guise of Smohalla exacting revenge on those who have built the dams and ruined the Columbia River for the salmon. For Charley, the final blow was the building of the Dalles Dam and destruction of Celilo Falls, which ironically is where Charley and Mary met. The backdrop to this tense situation is the rising waters on the Columbia and the question as to whether to draw down the river. The eventual drawdown is too little and too late.

Francine and Duke find distinctive harpoon tips in the dead bodies, but Francine doesn't want to turn them in to the authorities because she knows they can be linked to her father. Francine's feelings about the dams and their impact on the salmon are highly ambivalent. When she sees salmon feeding on the dead body at the hatchery, her reaction is mixed: "What was most chilling to Francine was that the image of salmon killing and eating their human destroyers was something she had fantasized about herself on many occasions. In her dream she was the murderer."[44] "Let's just keep this between us," Francine tells the hatchlings.[45]

Through a complicated series of events, Francine, Duke, and Jack meet (Francine and Duke fall in love), and Duke and Jack end up conspiring to create havoc on the Columbia. Jack goes into Canada, where he blows up the Mica Dam—the northernmost in the series of dams on the Columbia—thereby starting a cascade of flooding, dam breaking, and electrical failure. Duke ensures the destruction of Grand Coulee. But none of this is presented as a tragedy. Quite the contrary: almost no one is killed. The epilogue to *A River Out of Eden* is titled "Resurrection" and captures the rebirth of the Columbia:

Dam Failures, Real, Imagined, and Ecotage-Inspired   171

The creatures of a single mind became aware of the change far out to sea. It was a familiar sound [presumably the sound of the natural river], but they had never before heard it, and it drew them to places that the precious memory in their brains had nearly discarded. But when they became certain they were tasting the breath and soil of their ancient spawning grounds, they were suddenly only a single day away from those times. The river was speaking again to the sea. They would answer. Slowly they made their way from places far and near on the earth to the mouth of Nch'i-Wana (the mid-Columbia Indian name for the Columbia River).[46]

A joyful resurrection indeed and a positive example of ecotage.

Terrorism, ecoterrorism, sabotage, ecotage, or just a good deed—different people refer to the same act by different names, with very different biases and sympathies. As we have seen in this chapter, terrorism comes in many forms and with many motives. Dams excite hostility in certain circles—not necessarily against human beings but against gigantic earthen or concrete constructions regarded as dangerous and destructive, even malevolent. There will undoubtedly continue to be many terrorist dam novels (they often make for exciting reading), but one cannot help but wonder if a creeping sympathy for the fictional destroyers may not become more common in such works. The influence of Edward Abbey lives on, and the monkey wrench may be regarded as a positive tool by some determined environmentalists.

Conclusion

As we have seen, there was initially much enthusiasm for the building of big dams. In 1935, Franklin Delano Roosevelt gushed about the Hoover Dam on the Colorado River: "I came, I saw, and I was conquered." In 1948, Jawaharlal Nehru told skeptical villagers forced to move by a dam project: "If you are to suffer, you should suffer in the interest of the country." Sacrifice for dams makes sense, in other words. A few years later, Leonid Brezhnev declared this about the Bratsk Dam on Siberia's Angara River: "Today, where there was once dense taiga, the Bratsk hydroelectric station, the greatest in the world, has arisen." Behold the triumph of the Soviet system.[1]

Writers and intellectuals expressed themselves in glowing terms as well. Frank Waters called the Hoover Dam "the Great Pyramid of the American Desert, the Ninth Symphony of our day." Woody Guthrie announced that the Grand Coulee Dam on the Columbia River was "the mightiest thing ever built by a man." Anatolii Pristavkin wrote of the Angara: "She will go, my dear, darling Angara, a beautiful, wild, northern river, along a path that for

---

https://doi.org/10.5876/9781646425976.c006

now only exists on paper . . . houses, cities, settlements, a dam, industrial complexes, new factories . . . There will be everything, everything!"[2]

These sentiments changed. By the late 1950s, Nehru began to worry that "we are suffering from what we may call 'disease of gigantism.'" The Indian activist Baba Amte declared: "All those ruthless monstrous projects Nehru called temples of development have become tombs of development." The Indian writer Arundhati Roy stated bluntly: "Big Dams are to a nation's 'development' what nuclear bombs are to its military arsenal. They're both weapons of mass destruction." In 1984 Frank Waters also regretted his initial enthusiasm about the Hoover Dam: "Since then I have come to regard it as the first of our misguided attempts to dominate the entire natural world of the river."[3]

What happened? A sow's ear, it turns out, may not make a lovely silk purse. The initial expectation that dams would solve problems of flooding, irrigation, and water transport, plus provide pure and environmentally friendly hydroelectricity, was not always vindicated. Scouring below dams can contribute to worse flooding because of increased water flow. Overtopping of dams and their subsequent collapse can lead to many deaths and the destruction of land and structures. Evaporation increases salinization and dam reservoirs, it turns out, create additional atmospheric pollution in the form of greenhouse gases. The projected costs of dams often exceed expectations, as does their anticipated revenue. Reservoirs silt up, often faster than expected. Land near reservoirs is eroded, which contributes to worse flooding; animals suffer, and certain diseases abound. Reservoirs and river water become polluted, and fish populations decline in both quantity and quality. Undesirable creatures appear in reservoirs and in the silt behind them. Water may become not only un-potable but otherwise unusable. Salinization may lead to a decline, rather than a growth, in productive agriculture. Fears of terrorist attacks and dam-related earthquakes have paved the way to greater concerns about the possibility of ecotage, of human revenge on a modern, machine-dependent society for irremediable damage done to nature. The list goes on, and fiction attests to it in a lively and compelling way.

Most tragic, millions of people worldwide have suffered because of dams, often receiving inadequate financial compensation and being forced to move to poorer land and sometimes slums. Around 40 million to 60 million people have had to move because of dam construction. In most instances, the result has not been better lives for the typically already disadvantaged relocatees

174 CONCLUSION

but instead greater poverty, not to mention social alienation and a tragic sense of loss of their homeland and traditional existence. Migrant labor becomes more common, and a consequent, almost inevitable breakdown of social and family ties ensues. Women and the elderly in particular feel these strains. The recreational possibilities afforded by reservoirs cannot begin to compensate for these terrible losses.

This has led to an untenable situation that must be addressed, or there will be even more human suffering and negative environmental consequences. Decommissioning all or even most dams is not feasible, and decommissioning even a few is expensive and difficult. In *Cracked: The Future of Dams in a Hot, Chaotic World* (2023), Steven Hawley describes at length the complexities of dam removal.[4] Building fewer large dams in the future seems wise, and certainly fewer prime geological locations are available for mega-dams than there were a century ago. Financial greed, however, may interfere with more reasonable and informed intentions.

As the discussion in this study of the many literary works on the subject already produced has shown, mega-dams are not the great source of salvation they were once hoped to be. Rather, people suffer and the environment declines. It is not by chance that the longest chapter in this study concerns dispossession and alienation when people are forced to move so that dams can be built. Doubtless, more fiction and poetry will be written on this topic, as on environmental damage; ecotage may become a household word and a more common practice unless drastic attempts are made to remedy the situation. The general population will become more aware of the difficulties caused by dams, though, in part by reading the types of moving and dramatic fiction and poetry treated in this study. It is hoped that human beings will not persist in living in and creating a condition of deteriorating ecological decline but will instead attempt to repair what was done in the twentieth century more broadly and in a more organized fashion. Arrogant high modernism and the traditional popular focus on present rewards rather than the possibility of future calamities have led us to a crisis point. Only time will show whether restoration is possible, and literary works may well play an important role in determining the outcome. Certainly, authors of fiction are in an excellent position to represent dam-related problems skillfully as well as artistically, with an emphasis on the suffering of individual human beings and their attempts to assuage them.

Conclusion 175

Craig Lesley's and Elizabeth Cook-Lynn's Native Americans regret the formerly rich bounty of fish and the greater agricultural returns of the past. The elderly villagers of Valentin Rasputin's Matyora are emotionally distraught because they have to leave their beloved island, and Idris Ali's and Haggag Hassan Oddoul's Nubians long for their idyllic past on the Nile while living in Cairo slums. Li Mao Lovett's poor Chinese workers desperately scrabble though rubble looking for minimally valuable trash. Ugly reservoirs mar the Angara and the Volga, Nubian children suffer from a range of diseases that have invaded their environment, reservoirs are lined with barren land and dead trees, and pollution-laden sediment keeps building up inexorably and invisibly. As the literary works treated in this study often reveal in minute detail, what began with a conviction that a brave new world was going to emerge degenerated into a reality of poverty, devastation of the landscape, and social upheaval. Implicitly, these literary works demand a different and new approach to mega-dams, one that does not automatically assume that big dams will prevent flooding, improve agriculture, and promote electricity but purports that the building and care of dams should be conducted with much more thought and care than has been the case thus far.

Notes

Chapter 1: Introduction

1. Roosevelt's address is reprinted in its entirety in Dunar and McBride, *Building Hoover Dam*, 311.

2. Benjamin, *Invested Interests*, 107.

3. Owen, *Where the Water Goes*, 21.

4. Oscar Wilde, "The Decay of Lying," Virgil/.org/dswo/courses/novel/wilde-lying.pdf, 1.

5. Cited in Gadgil and Guha, *Ecology and Equity*, 185.

6. Cited in McNeill, *Something New under the Sun*, 149.

7. Dunar and McBride, *Building Hoover Dam*, 314.

8. Cited in Pearce, *The Dammed*, 219.

9. Spengler, *Decline of the West*, 501, original emphasis.

10. Mumford, *Technics and Civilization*, 37.

11. Cited in Adas, *Dominance by Design*, 143–44.

12. Lilienthal, *TVA*, 1.

13. Cited in Tsunts, *Rasskaz o Bol'shoi Volge*, 145.

14. Cited in Pearce, *The Dammed*, 133.

178 NOTES

15. Feshbach and Friendly, *Ecocide in the USSR*, 40.

16. Cullen, *Rivers in Harness*, 157.

17. Sinedubsky, *Power Giants on the Rivers*, 101.

18. Mauch and Zeller, "Rivers in History and Historiography," 3.

19. Pritchard, *Confluence*, 60.

20. Shapiro, *Mao's War against Nature*, 3–4.

21. Josephson, *Brute Force Technology*, 184–85.

22. Scott, *Seeing Like a State*, 4.

23. Dunar and McBride, *Building Hoover Dam*, 314.

24. Cf. Adas, *Dominance by Design*, 243.

25. Harvey, *Justice, Nature, and the Geography of Difference*, 121.

26. Zeisler-Vralsted, *Rivers, Memory, and Nation-Building*, 147.

27. Cf. Ekbladh, *Great American Mission*, 4.

28. Ekbladh, *Great American Mission*, 8.

29. Hawley, *Cracked*, 28.

30. Khagram, *Dams and Development*, 5.

31. Rist, *History of Development*, 39, original emphasis.

32. Weiss, Carayannis, Emmerij, and Jolly, *UN Voices*, 203.

33. Weiss, Carayannis, Emmerij, and Jolly, *UN Voices*, 217–18.

34. Amrith, *Unruly Waters*, 258.

35. Coates, *Story of Six Rivers*, 20.

36. Klingensmith, "One Valley and a Thousand," 2.

37. Josephson, *Brute Force Technology*, 25, 66.

38. Winchester, *River at the Center of the World*, 223.

39. Waterbury, *Hydropolitics of the Nile Valley*, 99.

40. Palmer, *Endangered Rivers*, 26.

41. Cited in Pearce, *The Dammed*, 81.

42. Cited in Fahim, *Dams, People, and Development*, 14.

43. Cited in McNeill, *Something New under the Sun*, 157. On Nehru's effusions, see also McCully, *Silenced Rivers*, 1–2.

44. Amrith, *Unruly Waters*, 201.

45. Palmer, *Endangered Rivers*, 31.

46. Grace, *Dam Nation*, 41.

47. Reisner, *Cadillac Desert*, 104.

48. On dam types, see, for example, Cullen, *Rivers in Harness*, 128–30; McCully, *Silenced Rivers*, 12–15; Turpin, *Dam*, 236–38.

49. Lilienthal, *TVA*, 53, original emphasis.

50. Rich, *Mortgaging the Earth*, 231.

51. Josephson, *Brute Force Technology*, 18.

52. Josephson, *Brute Force Technology*, 20.

53. Amrith, *Unruly Waters*, 202.

54. Reisner, *Cadillac Desert*, 158–59.

55. Cited in McNeill, *Something New under the Sun*, 159.

56. McCully, *Silenced Rivers*, 3–6.

57. Klingensmith, *"One Valley and a Thousand,"* 12.

58. There is a huge literature on relocation, both on the subject in general and specific case studies. Two of the most influential general works are Scudder, *Future of Large Dams*, and Goldsmith and Hildyard, *Social and Environmental Effects of Large Dams*. See also Adas, *Dominance by Design*; Ekbladh, *Great American Mission*; Everard, *Hydropolitics of Dams*; Josephson, *Brute Force Technology*; Khagram, *Dams and Development*; Leslie, *Deep Water*; McNeill, *Something New under the Sun*; Palmer, *Endangered Rivers and the Conservation Movement*; Parman, *Indians and the American West*; Pearce, *The Dammed*; Sneddon, *Concrete Revolution*.

59. Barber, *Death of Celilo Falls*, 152.

60. Josephson, *Brute Force Technology*, 53.

61. Adams, *Wasting the Rain*, 134.

62. Rich, *Mortgaging the Earth*, 156.

63. McNeill, *Something New under the Sun*, 159.

64. Pearce, *The Dammed*, 154.

65. Greener, *High Dam over Nubia*, 241.

66. Cited in Khagram, *Dams and Development*, 37.

67. Cited in Amrith, *Unruly Waters*, 211.

68. McCully, *Silenced Rivers*, 7–8.

69. Scudder, *Future of Large Dams*, 19.

70. Scudder, *Future of Large Dams*.

71. McCully, *Silenced Rivers*, 66.

72. McNeill, *Something New under the Sun*, 161.

73. Amrith, *Unruly Waters*, 295.

74. Rich, *Mortgaging the Earth*, 252.

75. On this problem, see especially McCully, *Silenced Rivers*, xxxii–xxxvii.

76. McCully, *Silenced Rivers*, xxxii.

77. McCully, *Silenced Rivers*, 87–92.

78. McCully, *Silenced Rivers*, 34.

79. McCully, *Silenced Rivers*, lii.

80. Khagram, *Dams and Development*, 179–81; McNeill, *Something New under the Sun*.

81. Brower, *"Foreword,"* 7.

82. McNeill, *Something New under the Sun*, 162.

180 NOTES

83. On the NBA (Narmada Bachao), see, for example, Amrith, *Unruly Waters*, 293–94; McNeill, *Something New under the Sun*, 162.

84. McCully, *Silenced Rivers*, lxviii.

85. Cited in Amrith, *Unruly Waters*, 213.

86. Waters, *The Colorado*, x.

87. Boulding, *Beasts, Ballads, and Bouldingisms*, 104.

Chapter 2: The High Modernist Heyday of Mega-Dam Construction

1. Some of the material presented in this chapter appeared in a different form and a different context, with a different purpose, in Ziolkowski, *Rivers in Russian Literature*, 49–59.

2. Trotskii, *Sochineniia*, 21:437, cited in Deutscher, *Prophet Unarmed*, 211.

3. Klingensmith, "One Valley and a Thousand," 213.

4. Gor'kii, *Po Soiuzu Sovetov*, 335.

5. Gor'kii, *Po Soiuzu Sovetov*, 339.

6. Gor'kii, *Po Soiuzu Sovetov*, 342.

7. Maiakovskii, "Dolg Ukraine," 7:229.

8. Bezymenskii, *Tragediinaia noch'*, 101.

9. Bezymenskii, *Tragediinaia noch'*, 104.

10. Bezymenskii, *Tragediinaia noch'*, 122.

11. Gladkov, *Energiia*, 2:47.

12. Gladkov, *Energiia*, 2:75.

13. Bash, *Goriachie chuvstva*, 108.

14. Bash, *Goriachie chuvstva*, 119.

15. Bash, *Goriachie chuvstva*, 303.

16. Bash, *Goriachie chuvstva*, 325.

17. Gonchar, *Chelovek i oruzhie*, 230–31.

18. Gonchar, *Chelovek i oruzhie*, 267.

19. Gor'kii, "Po Soiuzu Sovetov," 300.

20. In Ziolkowski, *Rivers in Russian Literature*, the discussion of the Russian literary treatment of dams on the Volga is found on pages 96–99.

21. Nesteruk, *Razvitie gidro-energetiki SSSR*, 115.

22. Nesteruk, *Razvitie gidro-energetiki SSSR*, 140.

23. Panferov, *Volga matushka-reka*, 1:11.

24. Panferov, *Volga matushka-reka*, 1:118.

25. Panferov, *Volga matushka-reka*, 1:240.

26. Panferov, *Volga matushka-reka*, 2:452.

27. Plashcheev and Chekmarev, *Gidrografiia SSSR*, 264.

28. St. George, *Siberia*, 147.

29. On early Soviet engineeering assessments of the Angara, see especially Ivanov, *Gidroenergetika Angary*, 13–17.

30. Cited in Buzhkevich, *Ot Paduna do Strelki*, 19.

31. Taurin, *Angara*, 1:18.

32. Taurin, *Angara*, 1:69.

33. Taurin, *Angara*, 1:72.

34. Taurin, *Angara*, 1:134.

35. Taurin, *Angara*, 1:492.

36. Taurin, *Angara*, 1:523.

37. Technical details about the Bratsk Hydroelectric Station are covered in Ivanov, *Gidroenergetika Angary*, 32–38.

38. Cited in Enisherlov and Ivanov, eds., *Eto bylo na Angare*, 9.

39. Buzhkevich, *Ot Paduna do Strelki*, 15.

40. Pristavkin, *Angara-reka*, 44.

41. Pristavkin, *Angara-reka*, 456.

42. Pristavkin, "Zapiski moego sovremennika," 140.

43. Evtushenko, *Bratskaia GES*, 180.

44. Evtushenko, *Bratskaia GES*, 238.

45. Extensive detail about the background and construction of Hoover Dam is provided in Billington and Jackson, *Big Dams of the New Deal Era*, 102–51.

46. For a detailed history of the planning and construction of Hoover Dam, see Hiltzik, *Colossus*.

47. Dunar and McBride, *Building Hoover Dam*, 306.

48. Waters, *The Colorado*, 329.

49. Gunther, *Inside U.S.A.*, 992.

50. Wilson, "Hoover Dam," 370.

51. Grey, *Boulder Dam*, 1.

52. Grey, *Boulder Dam*, 3.

53. On this aspect of Grey's thinking, see also Jackson, *Zane Grey*, 68.

54. Grey, *Boulder Dam*, 13.

55. Grey, *Boulder Dam*, 15.

56. Grey, *Boulder Dam*, 119.

57. Grey, *Boulder Dam*, 87–88.

58. Grey, *Boulder Dam*, 122.

59. Grey, *Boulder Dam*, 228.

60. Grey, *Boulder Dam*, 50.

61. Grey, *Boulder Dam*, 27.

182   NOTES

62. Grey, *Boulder Dam*, 109–110.

63. Haase, *Big Red*, 21.

64. Haase, *Big Red*, 275–76.

65. Haase, *Big Red*, 102.

66. Haase, *Big Red*, 182.

67. Haase, *Big Red*, 409.

68. Haase, *Big Red*, 142, original emphases.

69. Haase, *Big Red*, 37, 78, 154.

70. Haase, *Big Red*, 182.

71. Haase, *Big Red*, 112.

72. Haase, *Big Red*, 143.

73. Haase, *Big Red*, 256.

74. On dam construction on the Columbia, see Billington and Jackson, *Big Dams of the New Deal Era*, 152–99. On the Columbia River, see also White, *Organic Machine*.

75. Detailed accounts of Woody Guthrie's participation in this project are found in Cray, *Ramblin' Man*, 207–213; Kaufman, *Woody Guthrie's Modern World Blues*, 22–23; Alan Lomax's introduction to Murlin, *Woody Guthrie Roll on Columbia*, 7–9.

76. Lomax's introduction to Murlin, *Woody Guthrie Roll on Columbia*, 7.

77. Cited in Cray, *Ramblin' Man*, 209.

78. Cited in Cray, *Ramblin' Man*, 210.

79. Foreword to Cray, *Ramblin' Man*, xviii.

80. Harden, *A River Lost*, 89.

81. Lomax's introduction to Murlin, *Woody Guthrie Roll on Columbia*, 8.

82. Cited in Murlin, *Woody Guthrie Roll on Columbia*, 15.

83. Murlin, *Woody Guthrie Roll on Columbia*, 43.

84. Cited in Kaufman, *Woody Guthrie's Modern World Blues*, 23.

85. Sundborg, *Hail Columbia*, 430–31.

86. Gunther, *Inside U.S.A.*, 133.

87. Gunther, *Inside U.S.A.*, 131.

88. Gunther, *Inside U.S.A.*, 135.

89. On the TVA, see, for example, Chandler, *Myth of TVA*; Creese, *TVA's Public Planning*; McDonald and Muldowny, *TVA and the Dispossessed*.

90. Cited in McDonald and Muldowny, *TVA and the Dispossessed*, 263.

91. For a detailed history of the TVA, see especially Creese, *TVA's Public Planning*.

92. Verhoeven, *Water, Civilization, and Power in Sudan*, 139.

93. Buckles, *Valley of Power*, 54.

94. Buckles, *Valley of Power*, 78.

95. Buckles, *Valley of Power*, 10.

96. Buckles, *Valley of Power*, 100.

97. Buckles, *Valley of Power*, 58.

98. Buckles, *Valley of Power*, 245.

99. Buckles, *Valley of Power*, 208.

100. Deal, *Dunbar's Cove*, 7.

101. Deal, *Dunbar's Cove*, 90.

102. Deal, *Dunbar's Cove*, 82.

103. Deal, *Dunbar's Cove*, 140.

104. Deal, *Dunbar's Cove*, 338.

105. Deal, *Dunbar's Cove*, 419.

106. Gunther, *Inside U.S.A.*, 806.

107. Gunther, *Inside U.S.A*, 824.

Chapter 3: Displacement and Alienation of Peoples Worldwide

1. Pearce, *The Dammed*, 131.

2. Cf. McCully, *Silenced Rivers*, 101.

3. Palmer, *Endangered Rivers and the Conservation Movement*, 188.

4. Cf. Rubin and Warren, *Dams in Africa*, 31.

5. Goldsmith and Hildyard, *Social and Environmental Effects of Large Dams*, 44.

6. Rubin and Warren, *Dams in Africa*, 23–24. The authors go on to discuss the process in detail and with examples, 23–60.

7. Goldsmith and Hildyard, *Social and Environmental Effects of Large Dams*, 86.

8. Goldsmith and Hildyard, *Social and Environmental Effects of Large Dams*, 60.

9. Scudder, *The Future of Large Dams*, 20.

10. McCully, *Silenced Rivers*, 76.

11. Goldsmith and Hildyard, *Social and Environmental Effects of Large Dams*, 18.

12. Rich, *Mortgaging the Earth*, 155.

13. McCully, *Silenced Rivers*, 80–82.

14. Everard, *Hydropolitics of Dams*, 46

15. Scudder, *The Future of Large Dams*, 22–29.

16. On Rasputin's importance as a *derevenshchik*, see, for example, Gillespie, *Valentin Rasputin and Soviet Russian Village Prose*; Parthé, *Russian Village Prose*.

17. Rasputin, "*Vniz i verkh po techeniiu*," 1:493.

18. Rasputin, "*Vniz i verkh po techeniiu*," 1:494.

19. Valentin Rasputin, "*Proshchanie s Materoi*," 2:176.

20. Rasputin, "*Proshchanie s Materoi*," 2:266.

21. Rasputin, "*Proshchanie s Materoi*," 2:203.

22. Rasputin, "*Proshchanie s Materoi*," 2:234.

184 NOTES

23. Valentin Rasputin, *"Pozhar,"* 2:365.

24. Rasputin, *"Pozhar,"* 2:364.

25. Rasputin, *"Pozhar,"* 2:357.

26. Treaty with the Nez Perce, 1855, in Landeen and Pinkham, *Salmon and His People,* 236–37.

27. Extensive details on the lengthy saga of Indian fishing rights on the Columbia River, and at Celilo Falls in particular, are provided in Barber, *Death of Celilo Falls*: Parman, "Inconstant Advocacy"; Parman, *Indians and the American West*; Ulrich, *Empty Dams*. The situation of the Nez Perce in particular is outlined in Landham and Pinkham, *Salmon and His People*. On the Columbia River, see also White, *Organic Machine*.

28. Ulrich, *Empty Nets,* 71.

29. Ulrich, *Empty Nets,* 93.

30. Barber, *Death of Celilo Falls,* 7.

31. On Celilo Village, see Barber, *Death of Celilo Falls*.

32. Lesley, *Winterkill,* 5.

33. Lesley, *Winterkill,* 30.

34. Lesley, *Winterkill,* 69.

35. Lesley, *Winterkill,* 74.

36. Lesley, *Winterkill,* 192. Donald L. Parman makes a similar observation in "Inconstant Advocacy": "The Indians who had actually fished at Celilo Falls complained that their settlements did not adequately compensate them for the loss of the fishing rights. Almost instinctively, they sensed that the money would eventually be gone, while the fishing could have continued forever" (185).

37. Parman, "Inconstant Advocacy," 194.

38. Lesley, *River Song,* 162.

39. Lesley, *River Song,* 175.

40. Davies, "Euro-American Realism versus Native Authenticity," 241.

41. Lesley, *River Song,* 230.

42. Lesley, *River Song,* 274.

43. Lesley, *River Song,* 304.

44. Lawson, *Dammed Indians,* 1971.

45. Parman, *Indians and the American West,* 121.

46. The discussion here draws extensively on Lawson, *Dammed Indians,* especially 27–47, 135–36, 154–59.

47. Lawson, *Dammed Indians,* 29.

48. Cook-Lynn, *Aurelia,* 48.

49. Cook-Lynn, *Aurelia,* 49.

50. Cook-Lynn, *Aurelia,* 83.

51. Cook-Lynn, *Aurelia*, 129.

52. Cook-Lynn, *Aurelia*, 138–39.

53. Cook-Lynn, *Aurelia*, 178.

54. Cook-Lynn, *Aurelia*, 179.

55. Cook-Lynn, *Aurelia*, 185.

56. Cook-Lynn, *Aurelia*, 286.

57. Cook-Lynn, *Aurelia*, 439.

58. Cook-Lynn, *Aurelia*, 380.

59. Cook-Lynn, *Aurelia*, 441.

60. My discussion of the construction of the Aswan High Dam and the relocation of the Nubians draws on Adams, *Wasting the Rain*, 132–33; Collins, *The Nile*, 100–111, 120–21, 185–89; Collins, *Waters of the Nile*, 238–40, 267–68, 272–76; Dafalla, *Nubian Exodus*; Fahim, *Dams, People, and Development*, especially 26–27, 59–74; Fahim, *Egyptian Nubians*, especially 30–31, 115–16; Greener, *High Dam over Nubia*; Little, *High Dam at Aswan*; Said, *The River Nile*, especially 230–42; Waterbury, *Hydropolitics of the Nile Valley*, especially 98–109.

61. Collins, *Waters of the Nile*, 239.

62. Waterbury, *Hydropolitics of the Nile Valley*, 108.

63. On the construction of the Aswan High Dam, see Fahim, *Dams, People, and Development*.

64. Collins, *The Nile*, 185.

65. Fahim, *Dams, People, and Development*, 27.

66. Waterbury, *Hydropolitics of the Nile Valley*, 117.

67. Fahim, *Dams, People, and Development*, 72.

68. Fahim, *Egyptian Nubians*, 31

69. Abbas, "Egypt, Arab Nationalism, and Nubian Diasporic Identity," 151.

70. Ali, *Dongola*, 8.

71. Ali, *Dongola*, 20.

72. Ali, *Dongola*, 28–29.

73. Ali, *Dongola*, 71. Awad's failure to realize his dreams of a resurrected Nubia is discussed in DiMeo, "Unimaginable Community."

74. Ali, *Dongola*, 92.

75. Ali, *Poor*, 35–36.

76. Ali, *Poor*, 59.

77. Ali, *Poor*, 1.

78. Ali, *Poor*, 12.

79. Ali, *Poor*, 13.

80. Ali, *Poor*, 14.

81. Ali, *Poor*, 21.

186 NOTES

82. Ali, *Poor*, 93.

83. Ali, *Poor*, 96.

84. Ali, *Poor*, 105.

85. Ali, *Poor*, 112.

86. My discussion of dams in India draws on Amrith, *Unruly Waters*; Drèze, Samson, and Singh, *The Dam and the Nation*; D'Souza, *The Narmada Dammed*; Khagram, *Dams and Development*; Klingensmith, *"One Valley and a Thousand"*; Mathur, *Displacement and Resettlement in India*; McNeill, *Something New under the Sun*; Parasuraman, *The Development Dilemma*; Pearce, *The Dammed*; Rich, *Mortgaging the Earth*.

87. Khagram, *Dams and Development*, 35.

88. On the Scheduled Castes and Scheduled Tribes, see Smith, *Classifying the Universe*; Pail and Thoret, *Politics of Representation*.

89. Khagram, *Dams and Development*, 37.

90. Mathur, *Displacement and Resettlement in India*, 2.

91. D'Souza, *The Narmada Dammed*, xvi.

92. Parasuraman, *The Development Dilemma*, 83.

93. Parasuraman, *The Development Dilemma*, 93

94. Roy, "The Greater Common Good," in *The Cost of Living*, 80.

95. Cited in Pearce, *The Dammed*, 157.

96. Mirajkar, "Exploring the Full Potential of the Genre," 147.

97. "Reflection on Rural India," 2015.

98. Patil, *Dirge for the Dammed*, 4.

99. Patil, *Dirge for the Dammed*, 8.

100. Patil, *Dirge for the Dammed*, 9.

101. Patil, *Dirge for the Dammed*, 26.

102. Patil, *Dirge for the Dammed*, 26.

103. Patil, *Dirge for the Dammed*, 34.

104. Patil, *Dirge for the Dammed*, 55.

105. Patil, *Dirge for the Dammed*, 114.

106. Patil, *Dirge for the Dammed*, 240.

107. Patil, *Dirge for the Dammed*, 231.

108. Patil, *Dirge for the Dammed*, 458.

109. Patil, *Dirge for the Dammed*, 466.

110. Patil, *Dirge for the Dammed*, 471.

111. My discussion of dams in China draws on Khagram, *Dams and Development*, especially 172–73; Murray and Cook, *Green China*; Pearce, *The Dammed*, especially 237–41; Shapiro, *Mao's War against Nature*; Winchester, *The River at the Center of the World*; Van Slyke, *Yangtze*.

112. Murray and Cook, *Green China*, 99; Van Slyke, *Yangtze*, 183.

Notes 187

113. On the Three Gorges Dam, see especially Khagram, *Dams and Developments*; Pearce, *The Dammed*; Shapiro, *Mao's War against Nature*.

114. Cited in Khagram, *Dams and Development*, 172–73.

115. Winchester, *The River at the Center of the World*, 223.

116. Shapiro, *Mao's War against Nature*, 14.

117. Murray and Cook, *Green China*, 102–3.

118. Lovett, *In the Lap of the Gods*, 9.

119. Lovett, *In the Lap of the Gods*, 298.

120. Lovett, *In the Lap of the Gods*, 52.

121. Lovett, *In the Lap of the Gods*, 52.

122. Lovett, *In the Lap of the Gods*, 110–11.

Chapter 4: Contaminated Water, Disappearing Fish, and Deadly Sediment

1. Everard, *Hydropolitics of Dams*, 133.

2. Goldsmith and Hildyard, *Social and Environmental Effects of Large Dams*, 51–231; McCully, *Silenced Rivers*, 29–64.

3. Goldsmith and Hildyard, *Social and Environmental Effects of Large Dams*, 219.

4. McCully, *Silenced Rivers*, 29–30.

5. McCully, *Silenced Rivers*, 41.

6. McCully, *Silenced Rivers*, 58.

7. Sneddon, *Concrete Revolution*, 142–43.

8. My discussion of environmental damage to Russian rivers draws on the analysis found in Ziolkowski, *Rivers in Russian Literature*, 99–100, 193–97.

9. Feshbach and Friendly, *Ecocide in the USSR*, 98.

10. Rasputin, *"Vniz i verkh po techeniiu,"* 1:478.

11. Rasputin, *"Vniz i verkh po techeniiu,"* 496.

12. Rasputin, *"Vniz i verkh po techeniiu,"* 502.

13. Rasputin, *"Vniz i verkh po techeniiu,"* 504.

14. Chudakova, *"Plyvushchii korabl',"* 231.

15. Viktor Drobotov, *"Stony Volgi,"* 42.

16. Drobotov, *"Stony Volgi,"* 124.

17. N. E. Pal'kin, *"O Volge s nadezhdoi I bol'iu,"* 219.

18. Pal'kin, *"O Volge s nadezhdoi i bol'iu,"* 238.

19. Lesley, *River Song*, 49.

20. Lesley, *River Song*, 74.

21. Lesley, *River Song*, 76.

NOTES

22. Gulick, *Roll on, Columbia*, bk. 3: *Into the Desert*, 336.

23. Gulick, *The Dam Builders*, 53.

24. On hatcheries and their difficulties, see Hawley, *Cracked*, 183–87.

25. Gulick, *The Dam Builders*; Hawley, *Cracked*, 330.

26. Gulick, *The Dam Builders*; Hawley, *Cracked*, 351.

27. Cook-Lynn, *Aurelia*, 164–65.

28. Cook-Lynn, *Aurelia*, 172, 287, 380, 422.

29. Cook-Lynn, *Aurelia*, 324.

30. Cook-Lynn, *Aurelia*, 351.

31. Cook-Lynn, *Aurelia*, 437–38.

32. Cook-Lynn, *Aurelia*, 458.

33. Ali, *Dongola*, 23.

34. Ali, *Dongola*, 23–24.

35. Ali, *Dongola*, 38.

36. Ali, *Poor*, 43.

37. Ali, *Poor*, 81.

38. Oddoul, "Adila, Grandmother," in *Nights of Musk: Stories from Old Nubia*, 11.

39. Oddoul, "The River People," in *Nights of Musk*, 93–94, 96.

40. Cited in Oddoul, "The River People," 54.

41. McCully, *Silenced Rivers*, 54.

42. Patil, *Dirge for the Dammed*, 8.

43. Patil, *Dirge for the Dammed*, 38.

44. Patil, *Dirge for the Dammed*, 197.

45. Patil, *Dirge for the Dammed*, 199.

46. McCully, *Silenced Rivers*, 107.

47. Goldsmith and Hildyard, *Social and Environmental Effects of Large Dams*, 221.

48. McCully, *Silenced Rivers*, 109.

49. Everard, *Hydropolitics of Dams*, 63.

50. McCully, *Silenced Rivers*, 145.

51. McCully, *Silenced Rivers*, 110.

52. Goldsmith and Hildyard, *Social and Environmental Effects of Large Dams*, 105.

53. Goldsmith and Hildyard, *Social and Environmental Effects of Large Dams*, 110–12.

54. McCully, *Silenced Rivers*, 112.

55. Leopold, "Sediment Problems at the Three Gorges Dam," 195.

56. See, for example, Everard, *Hydropolitics of Dams*, 52.

57. Armstrong, *Taming the Dragon*, vii.

58. A detailed account of John Hersey's career and concerns is found in Sanders, *John Hersey Revisited*. Sanders calls *A Single Pebble* a "story of an engineer who failed as an evangelist for Western progress" (48).

59. Hersey, *A Single Pebble*, 4.

60. Hersey, *A Single Pebble*, 107–8.

61. Hersey, *A Single Pebble*, 111, original emphasis.

62. Hersey, *A Single Pebble*, 180–81.

63. Van Slyke, *Yangtze*, 182.

64. Armstrong, *Taming the Dragon*, 1.

65. Armstrong, *Taming the Dragon*, 2.

66. Armstrong, *Taming the Dragon*, 3.

67. Armstrong, *Taming the Dragon*, 4.

68. Armstrong, *Taming the Dragon*, 5.

69. On this point, see, for example, Chetham, *Before the Deluge*, 185.

70. Armstrong, *Taming the Dragon*, 18.

71. Armstrong, *Taming the Dragon*, 27.

72. Armstrong, *Taming the Dragon*, 31.

73. Armstrong, *Taming the Dragon*, 40.

74. Armstrong, *Taming the Dragon*, 43.

75. Armstrong, *Taming the Dragon*, 44.

76. Armstrong, *Taming the Dragon*, 52.

77. Chetham, *Before the Deluge*, 1–2.

78. Chetham, *Before the Deluge*, 54.

79. Chetham, *Before the Deluge*, 66.

80. Chetham, *Before the Deluge*, 159.

81. Chetham, *Before the Deluge*, vii.

82. Kuo, *The Man Who Dammed the Yangtze*, 33.

83. Kuo, *The Man Who Dammed the Yangtze*, 80.

84. Kuo, *The Man Who Dammed the Yangtze*, 100.

85. Kuo, *The Man Who Dammed the Yangtze*, 127.

86. Kuo, *The Man Who Dammed the Yangtze*, 115.

87. Kuo, *The Man Who Dammed the Yangtze*, 142.

88. Kuo, *The Man Who Dammed the Yangtze*, 162.

89. Kuo, *The Man Who Dammed the Yangtze*, 179.

90. Xiaolong, *Don't Cry, Tai Lake*, 105.

Chapter 5: Dam Failures, Real, Imagined, and Ecotage-Inspired

1. On the history of the Johnstown Flood, see McCullough, *The Johnstown Flood*.

2. On the St. Francis Dam, see Outland, *Man-made Disaster*. For a general overview of dam failures, see McCully, *Silenced Rivers*, 101–32.

190 NOTES

3. Hastings, *Operation Chastise*.

4. My account of the events surrounding the Johnstown Flood relies primarily on McCullough, *The Johnstown Flood*.

5. McCullough, *The Johnstown Flood*, 258.

6. McCullough, *The Johnstown Flood*, 262.

7. Cambor, *In Sunlight*, 84.

8. Cambor, *In Sunlight*, 242.

9. Cambor, *In Sunlight*, 243.

10. Coopey, *Waterproof*, 159.

11. Coopey, *Waterproof*, 153.

12. Coopey, *Waterproof*, 159.

13. On the Aswan High Dam, see Collins, *The Nile*; Fahim, *Dams, People, and Development*.

14. Heim, *Aswan*, 46.

15. Heim, *Aswan*, 48.

16. Heim, *Aswan*, 275.

17. Harrington and Young, *Aswan High*, 41.

18. Harrington and Young, *Aswan High*, 229.

19. On Muir's opposition to Hetch Hetchy, see Scarce, *Eco-warriors*.

20. On this famous controversy, see Scarce, *Eco-warriors*, 17–19.

21. On Echo Park, Dinosaur National Monument, and Glen Canyon, see, for example, McCully, *Silenced Rivers*, 283–84.

22. Abbey, "Down the River," 134.

23. Abbey, "Down the River," 188.

24. Likar, *Eco-warriors*, 4.

25. Likar, *Eco-warriors*, 4.

26. Cf. Likar, *Eco-warriors*, 5, 35.

27. Likar, *Eco-warriors*, 119.

28. Likar, *Eco-warriors*, 85.

29. Calahan, *Edward Abbey*, 161.

30. Likar, *Eco-warriors*, 104.

31. Likar, *Eco-warriors*, 85.

32. Scarce, *Eco-warriors*, 12.

33. Abbey, *The Monkey Wrench Gang*, 32.

34. Abbey, *The Monkey Wrench Gang*, 5–6.

35. Quigley, "Introduction," 6.

36. Hansen, *Wet Desert*, 5.

37. Hansen, *Wet Desert*, 13.

38. Hansen, *Wet Desert*, 19.

39. Hansen, *Wet Desert*, 301.

40. Moody, *K Falls*, i.

41. Moody, *K Falls*, ii.

42. Moody, *K Falls*, 3

43. Moody, *K Falls*, 199.

44. Hockenberry, *River out of Eden*, 13.

45. Hockenberry, *River out of Eden*, 15.

46. Hockenberry, *River out of Eden*, 361.

Conclusion

1. Roosevelt quoted in Dunar and McBride, *Building Hoover Dam*, 311; Nehru quoted in Khagram, *Dams and Development*, 37; Brezhnev quoted in Enisherlov and Ivanov, *Eto bylo na Angare*, 9.

2. Waters, *The Colorado*, 329; Guthrie quoted in Murlin, *Woody Guthrie Roll on Columbia*, 15; Pristavkin, *Angara-reka*, 456.

3. Nehru quoted in Amrith, *Unruly Waters*, 213; Amte quoted in Pearce, *The Dammed*, 157; Roy, "The Greater Common Good," 80; Waters, *The Colorado*, x.

4. Hawley, *Cracked*, 191–21.

Bibliography

Abbas, Fatim. "Egypt, Arab Nationalism, and Nubian Diasporic Identity in Idris Ali's *Dongola: A Novel of Nubia*." *Research in African Literature* 45 (2014): 147–66.

Abbey, Edward. *Desert Solitaire*. Tucson: University of Arizona Press, 1968.

Abbey, Edward. "Down the River." In *Desert Solitaire*, by Abbey, 134–172. Tucson: University of Arizona Press, 1968.

Abbey, Edward. *The Monkey Wrench Gang*. New York: HarperPerennial Modern Classics, 2000 [1975].

Adams, W. M. *Wasting the Rain: Rivers, People, and Planning in Africa*. Minneapolis: University of Minnesota Press, 1992.

Adas, Michael. *Dominance by Design: Technological Imperatives and America's Civilizing Mission*. Cambridge, MA: Belknap Press of Harvard University Press, 2006.

Ali, Idris. *Dongola: A Novel of Nubia*. Translated by Peter Theroux. Fayetteville: University of Arkansas Press, 1998.

Ali, Idris. *Poor*. Translated by Elliott Colla. Cairo: American University in Cairo Press, 2007.

Amrith, Sunil. *Unruly Waters: How Rains, Rivers, Coasts, and Seas Have Shaped Asia's History*. New York: Basic Books, 2018.

Armstrong, Dan. *Taming the Dragon*. New York: iUniverse, 2007.

https://doi.org/10.5876/9781646425976.c007

BIBLIOGRAPHY

Barber, Katrine. *Death of Celilo Falls*. Seattle: University of Washington Press, 2005.

Bash, Iakov. *Goriachie chuvstva*. Translated by Vladimir Iurezanskii. Moscow: Izdatel'stvo VtsSPS Profizdat, 1951.

Benjamin, Bret. *Invested Interests: Capital, Culture, and the World Bank*. Minneapolis: University of Minnesota Press, 2007.

Bezymenskii, Aleksandr. *Tragediinaia noch': Poema*. Moscow: Sovetskii pisatel', 1964.

Billington, David P., and Donald C. Jackson. *Big Dams of the New Deal Era: A Confluence of Engineering and Politics*. Norman: University of Oklahoma Press, 2006.

Boulding, Kenneth E. *Beasts, Ballads, and Bouldingisms*. Edited by Richard P. Beilock. New Brunswick, NJ: Transaction Books, 1980.

Brower, David Ross. "Foreword." In Eliot Porter, photographer, *The Place No One Knew: Glen Canyon*, edited by David Ross Brower, 8–10. San Francisco: Sierra Club, 1963.

Buckles, Eleanor. *Valley of Power*. New York: Creative Age Press, 1945.

Buzhkevich, Miroslav. *Ot Paduna do Strelki*. Moscow: Mysl', 1965.

Calahan, James M. *Edward Abbey: A Life*. Tucson: University of Arizona Press, 2001.

Cambor, Kathleen. *In Sunlight, in a Beautiful Garden*. New York: Farrar, Straus, and Giroux, 2001.

Chandler, William U. *The Myth of TVA: Conservation and Development in the Tennessee Valley, 1933–1983*. Cambridge, MA: Ballinger, 1984.

Chetham, Deirdre. *Before the Deluge: The Vanishing World of the Yangtze's Three Gorges*. New York: Palgrave Macmillan, 2002.

Chudakova, Marietta. "*Plyvushchii korabl'*." *Novyi mir* 7 (1989): 231–35.

Coates, Peter. *A Story of Six Rivers: History, Culture, and Ecology*. London: Reaktion Books, 2013.

Collins, Robert O. *The Nile*. New Haven, CT: Yale University Press, 2002.

Collins, Robert O. *The Waters of the Nile: Hydropolitics and the Jonglei Canal, 1900–1988*. Oxford: Oxford University Press, 1990.

Cook-Lynn, Elizabeth. *Aurelia: A Crow Creek Trilogy*. Niwot: University Press of Colorado, 1999.

Coopey, Judith Redline. *Waterproof: A Novel of the Johnstown Flood*. Mesa, AZ: Fox Hollow, 2011.

Cray, Ed. *Ramblin' Man: The Life and Times of Woody Guthrie*. New York: W. W. Norton, 2004.

Creese, Walter L. *TVA's Public Planning: The Vision, the Reality*. Knoxville: University of Tennessee Press, 1990.

Cullen, Allen H. *Rivers in Harness: The Story of Dams*. Philadelphia: Chilton Books, 1962.

Dafalla, Hassan. *The Nubian Exodus*. London: C. Hurst, 1975.

Davies, J. C. "Euro-American Realism versus Native Authenticity: Two Novels by Craig Lesley." *Studies in American Fiction* 22 (1994): 234–47.

Deal, Borden. *Dunbar's Cove*. London: Hutchinson, 1958.

Deutscher, Isaac. *The Prophet Unarmed: Trotsky, 1921–1929.* Oxford: Oxford University Press, 1959.

DiMeo, David. "Unimaginable Community: The Failure of Nubian Nationalism in Idris Ali's *Dongola*." *Research in African Literature* 46 (2015): 72–89.

Drèze, Jean, Meera Samson, and Satyajit Singh, eds. *The Dam and the Nation: Displacement and Resettlement in the Narmada Valley.* Delhi: Oxford University Press, 1997.

Drobotov, Victor. "*Stony Volgi.*" In *Stony Volgi*, edited by A. V. Kokshilov, 6–42. Volgograd: Nizhne-volzhkoe knizhnoe izdatel'stvo, 1990.

D'Souza, Dilip. *The Narmada Dammed: An Inquiry into the Politics of Development.* New Delhi: Penguin Books, 2002.

Dunar, Andrew J., and Dennis McBride, eds. *Building Hoover Dam: An Oral History of the Great Depression.* Reno: University of Nevada Press, 1993.

Ekbladh, David. *The Great American Mission: Modernization and the Construction of an American World Order.* Princeton, NJ: Princeton University Press, 2010.

Enisherlov, V., and D. Ivanov, eds. *Eto bylo na Angare.* Moscow: Molodaia gvardiia, 1974.

Everard, Mark. *The Hydropolitics of Dams: Engineering or Ecosystems?* London: Zed Books, 2013.

Evtushenko, Evgenii. *Bratskaia GESI: Stikhi i poema.* Moscow: Sovetskii pisatel', 1967.

Fahim, Hussein M. *Dams, People, and Development: The Aswan High Dam Case.* New York: Pergamon, 1981.

Fahim, Hussein M. *Egyptian Nubians: Resettlement and Years of Coping.* Salt Lake City: University of Utah Press, 1983.

Feshbach, Murray, and Alfred Friendly Jr. *Ecocide in the USSR: Health and Nature under Siege.* New York: Basic Books, 1992.

Gadgil, Madhav, and Ramachandra Guha. *Ecology and Equity: The Use and Abuse of Nature in Contemporary India.* London: Routledge, 1995.

Gillespie, David C. *Valentin Rasputin and Soviet Russian Village Prose.* London: MHRH, 1986.

Gladkov, Fedor. *Sobranie sochinenii.* 5 vols. Moscow: Khudozhestvennnaia literatura, 1983–85.

Goldsmith, Edward, and Nicholas Hildyard. *The Social and Environmental Effects of Large Dams.* San Francisco, CA: Sierra Club Books, 1984.

Gonchar, Oles'. *Chelovek i oruzhie.* Translated by M. Alekseev and I. Karabatenko. Moscow: Molodaia gvardiia, 1960.

Gor'kii, Maksim. *Delo Artamonovykh: Po Soiuzu Sovetov, V. I. Lenin.* Moscow: Izdatel'stvo Izvestiia, 1964.

Grace, Stephen. *Dam Nation: How Water Shaped the West and Will Determine Its Future.* Guilford, CT: Globe Pequot, 2012.

Greener, Leslie. *High Dam over Nubia.* New York: Viking, 1962.

Grey, Zane. *Boulder Dam.* Roslyn, NY: Walter J. Black, 1963.

196 BIBLIOGRAPHY

Gulick, Bill. *The Dam Builders*. Albuquerque: University of New Mexico Press, 2008.

Gulick, Bill. *Roll on, Columbia*, bk. 3: *Into the Desert*. Niwot: University Press of Colorado, 1998.

Gunther, John. *Inside U.S.A.* Revised edition. New York: Harper and Brothers, 1947.

Haase, John. *Big Red*. New York: Harper and Row, 1980 [1966].

Hansen, Gary. *Wet Desert: Tracking down a Terrorist on the Colorado River*. Kaysville, UT: Hole Shot Press, 2007.

Harden, Blaine. *A River Lost: The Life and Death of the Columbia*. New York: W. W. Norton, 2012 [1996].

Harrington, R. E., and James A. Young. *Aswan High*. London: Secker and Warburg, 1983.

Harvey, David. *Justice, Nature, and the Geography of Difference*. Cambridge, MA: Blackwell, 1996.

Hastings, Max. *Operation Chastise*. New York: Harper, 2020.

Hawley, Steven. *Cracked: The Future of Dams in a Hot, Chaotic World*. Ventura, CA: Patagonia Works, 2023.

Heim, Michael. *Aswan!* Translated by J. Maxwell Brownjohn. New York: Alfred A. Knopf, 1972.

Hersey, John. *A Single Pebble*. New York: Alfred A. Knopf, 1963.

Hiltzich, Michael. *Colossus: Hoover Dam and the Making of the American Century*. New York: Free Press, 2010.

Hockenberry, John. *A River Out of Eden*. New York: Doubleday, 2001.

Ivanov, I. I. *Gidroenergetika Angary i prirodnaia sreda*. Novosibirsk, Russia: Nauka, 1991.

Jackson, Carlton. *Zane Grey*. Revised edition. Boston: Twayne, 1968.

Jackson, Donald C. *Pastoral and Monumental: Dams, Postcards, and the American Landscape*. Pittsburgh, PA: University of Pittsburgh Press, 2013.

Josephson, Paul R. *Brute Force Technology and the Transformation of the Natural World*. Washington, DC: Island, 2002.

Kaufman, Will. *Woody Guthrie's Modern World Blues*. Norman: University of Oklahoma Press, 2017.

Khagram, Sanjeev. *Dams and Development: Transnational Struggles for Water and Power*. Ithaca, NY: Cornell University Press, 2004.

Klingensmith, Daniel. *"One Valley and a Thousand": Dams, Nationalism, and Development*. New Delhi: Oxford University Press, 2007.

Kokshilov, A. V., ed. *Stony Volgi*. Volgograd: Nizhne-volzhkoe knizhnoe izdatel'stvo, 1990.

Kuo, Alex. *The Man Who Dammed the Yangtze*. Hong Kong: Haven Books, 2011.

Landeen, Dan, and Allen Pinkham. *Salmon and His People: Fish and Fishing in the Nez Perce Culture*. Lewiston, ID: Confluence, 1999.

Lawson, Michael L. *Dammed Indians: The Pick-Sloan Plan and the Missouri River Sioux, 1944–1980*. Norman: University of Oklahoma Press, 1971.

Lemeshev, M. Ia., ed. *Ekologicheskaia al'ternativa*. Moscow: Progress, 1990.

Leopold, Luna B. "Sediment Problems at the Three Gorges Dam." In *The River Dragon Has Come! The Three Gorges Dam and the Fate of China's Yangtze River and Its People*, edited by Dai Qing, John G. Thibodeau, and Philip B. Williams, 194–99. Translated by Yi Ming. Armonk, NY: M. E. Sharpe, 1998.

Lesley, Craig. *River Song*. Boston: Houghton Mifflin, 1989.

Lesley, Craig. *Winterkill*. New York: Dell, 1984.

Leslie, Jacques. *Deep Water: The Epic Struggle over Dams, Displaced People, and the Environment*. New York: Farrar, Straus and Giroux, 2005.

Likar, Lawrence E. *Eco-warriors, Nihilistic Terrorists, and the Environment*. Santa Barbara, CA: Praeger, 2011.

Lilienthal, David E. *TVA: Democracy on the March*. Chicago: Quadrangle Books, 1966 [1941].

Little, Tom. *High Dam at Aswan: The Subjugation of the Nile*. New York: John Day, 1965.

Lovett, Li Mao. *In the Lap of the Gods*. Teaticket, MA: Leapfrog, 2010.

Maiakovskii, Vladimir. *Polnoe sobranie sochinenii*. 13 vols. Moscow: Gosudarstvennoe izdatel'stvo khudozhestvennoi literatury, 1955–61.

Mathur, Hari Mohan. *Displacement and Resettlement in India: The Human Cost of Development*. London: Routledge, 2013.

Mauch, Christof, and Thomas Zeller. "Rivers in History and Historiography: An Introduction." In *Rivers in History: Perspectives on Waterways in Europe and North America*, edited by Christof Mauch and Thomas Zeller, 1–10. Pittsburgh, PA: University of Pittsburgh Press, 2008.

McCullough, David. *The Johnstown Flood*. London: Hutchinson, 1968,

McCully, Patrick. *Silenced Rivers: The Ecology and Politics of Large Dams*. London: Zed Books, 2007.

McDonald, Michael J., and John Muldowny. *TVA and the Dispossessed: The Resettlement of Population in the Norris Dam Area*. Knoxville: University of Tennessee Press.

McNeill, J. R. *Something New under the Sun: An Environmental History of the Twentieth-Century World*. New York: W. W. Norton, 2000.

Michaels, Anne. *The Winter Vault*. New York: Alfred A. Knopf, 2009.

Mirajkar, Nishikant D. "Exploring the Full Potential of the Genre." *Indian Literature* 37 (1994): 147–54.

Moody, Skye Kathleen. *K Falls*. New York: St. Martin's Minotaur, 2001.

Mumford, Lewis. *Technics and Civilization*. New York: Harcourt, Brace and World, 1963.

Murlin, Bill, ed. *Woody Guthrie Roll on Columbia: The Columbia River Collection*. Bethlehem, PA: Sing Out Corporation, 1991.

Murray, Geoffrey, and Ian G. Cook. *Green China: Seeking Ecological Alternatives*. New York: Routledge Curzon, 2002.

BIBLIOGRAPHY

Nesteruk, F. Ia. *Razvitie gidro-energetiki SSSR*. Moscow: Izdatel'stvo Akademii nauk SSSSR, 1963.

Oddoul, Haggag Hassan. "Adila, Grandmother." In *Nights of Musk: Stories from Old Nubia*, 1–28. Translated by Anthony Calderbank. Cairo: American University in Cairo Press, 2005.

Oddoul, Haggag Hassan. *Nights of Musk: Stories from Old Nubia*. Translated by Anthony Calderbank. Cairo: American University in Cairo Press, 2005.

Oddoul, Haggag Hassan. "The River People." In *Nights of Musk*, 89–126. Translated by Anthony Calderbank. Cairo: American University in Cairo Press, 2005.

Outland, Charles F. *Man-made Disaster: The Story of the St. Francis Dam*. Glendale, CA: A. H. Clark, 1977.

Owen, David. *Where the Water Goes: Life and Death along the Colorado River*. New York: Riverhead Books, 2017.

Pai, Sudha, and Sukhadeo Thorat, eds. *Politics of Representation: Historically Disadvantaged Groups in India's Democracy*. Singapore: Palgrave Macmillan, 2022.

Pal'kin, N. E. "O Volge s nadezhdoi i bol'iu." In *Ekologicheskaia al'ternativa*, edited by M. Ia. Lemeshev, 217–38. Moscow: Profress, 1990.

Palmer, Tim. *Endangered Rivers and the Conservation Movement*. 2nd ed. Lanham, MD: Rowman and Littlefield, 2004.

Panferov, Fedor. *Volga matushka-reka*. 2 vols. Moscow: Sovetskaia Rossiia, 1958.

Parasuraman, S. *The Development Dilemma: Displacement in India*. New York: St. Martin's, 1999.

Parman, Donald L. "Inconstant Advocacy: The Erosion of Indian Fishing Rights in the Pacific Northwest, 1933–1956." *Pacific Historical Review* 53 (1984): 163–89.

Parman, Donald L. *Indians and the American West in the Twentieth Century*. Bloomington: Indiana University Press, 1994.

Parthé, Kathleen F. *Russian Village Prose: The Radiant Past*. Princeton, NJ: Princeton University Press, 1992.

Patil, Vishwas. *A Dirge for the Dammed*. Translated by Keerti Ramachandra. Gurgaon, India: Hachette India, 2014.

Pearce, Fred. *The Dammed: Rivers, Dams, and the Coming World Water Crisis*. London: Bodley Head, 1992.

Plashcheev, A. V., and V. A. Chekmarev. *Gidrografiia SSSR*. Leningrad: Gidrometeorologicheskoe izdatel'stvo, 1967.

Pristavkin, Anatolii. *Angara-reka*. Moscow: PROFIZDAT, 1977.

Pristavkin, Anatolii. "Zapiski moego sovremennika." In *Sibirskie povesti*, edited by Anatolii Pristavkin, 69–239. Novosibirsk, Russia: Zapadno-sibirskoe knizhnoe izdatel'stvo, 1967.

Pritchard, Sara B. *Confluence: The Nature of Technology and the Remaking of the Rhône*. Cambridge, MA: Harvard University Press, 2011.

Qing, Dai, ed. *The River Dragon Has Come!* Edited by John G. Thibodeau and Philip B. William. Translated by Yi Ming. Armonk, NY: M. E. Sharpe, 1998.

Quigley, Peter, ed. *Coyote in the Maze: Tracking Edward Abbey in a World of Words*. Salt Lake City: University of Utah Press, 1998.

Quigley, Peter. "Introduction: Heraclitean Fire, Bakhtinian Laughter, and the Limits of Literary Judgment." In *Coyote in the Maze*, by Peter Quigley, 1–18. Salt Lake City: University of Utah Press, 1998.

Rasputin, Valentin. *"Pozhar."* In *Sobranie sochinenii*, by Rasputin, 2:365. Moscow: Molodaia gvardiia, 1994.

Rasputin, Valentin. *"Proshchanie s Materoi."* In *Sobranie sochinenii*, by Rasputin, 2:176. Moscow: Molodaia gvardiia, 1994.

Rasputin, Valentin. *Sobranie sochinenii*. 3 vols. Moscow: Molodaia gvardiia, 1994.

Rasputin, Valentin. *"Vniz i verkh po techeniiu."* In *Sobranie sochinenii*, by Rasputin, 1:493. Moscow: Molodaia gvardiia, 1994.

"A Reflection on Rural India." June 25, 2015. https://www.domain-b.com/people/interviews/20150625_vishwas_patil.html.

Reisner, Marc. *Cadillac Desert: The American West and Its Disappearing Water*. New York: Penguin Books, 1993.

Rich, Bruce. *Mortgaging the Earth: The World Bank, Environmental Impoverishment, and the Crisis of Development*. Boston: Beacon, 1994.

Rist, Gilbert. *The History of Development: From Western Origins to Global Faith*. 3rd ed. Translated by Patrick Camiller. London: Zed Books, 2008.

Rowe, John, *The Aswan Solution*. Garden City, NY: Doubleday, 1979.

Roy, Arundhati. *The Cost of Living*. New York: Modern Library, 1999.

Rubin, Neville, and William M. Warren, eds. *Dams in Africa: An Inter-disciplinary Study of Man-made Lakes in Africa*. London: Frank Cass, 1968.

Said, Rushdi. *The River Nile: Geology, Hydrology, and Utilization*. Oxford: Pergamon, 1993.

Sanders, David. *John Hersey Revisited*. Boston: Twayne, 1990.

Scarce, Rik. *Eco-warriors: Understanding the Radical Environmental Movement*. Walnut Creek, CA: Left Coast Press, 2006 [1990].

Scott, James C. *Seeing Like a State: How Certain Schemes to Improve the Human Condition Have Failed*. New Haven, CT: Yale University Press, 1998.

Scudder, Thayer. *The Future of Large Dams: Dealing with Social, Environmental, Institutional, and Political Costs*. London: Earthscan, 2005.

Serbskii, Viktor Solomonovich, and Ekaterina Viktorovna Serbskaia, eds. *Veter Bratska: Sbornik stikhov*. Bratsk: Bratskaia gorodskaia tipografiia, 1995.

Shapiro, Judith. *Mao's War against Nature: Politics and the Environment in Revolutionary China*. Cambridge: Cambridge University Press, 2001.

Sinedubsky, Vladimir. *Power Giants on the Rivers*. Moscow: Novosti, 1965.

Smith, Brian K. *Clarifying the Universe: The Ancient Indian Varna System and the Origins of Caste*. New York: Oxford University Press, 1994.

Sneddon, Christopher. *Concrete Revolution: Large Dams, Cold War Geopolitics, and the US Bureau of Reclamation*. Chicago: University of Chicago Press, 2015.

200 BIBLIOGRAPHY

Spengler, Oswald. *The Decline of the West*, vol. 2: *Perspectives of World History*. Translated by Charles Granise Atkinson. New York: Knopf, 1939.

St. George, George. *Siberia: The New Frontier*. New York: David McKay, 1969.

Sundborg, George. *Hail Columbia: The Thirty-Year Struggle for Grand Coulee Dam*. New York: Macmillan, 1954.

Taurin, Frants. *Angara: Roman*. In *Izbrannye proizvedeniia*. 2 vols. Moscow: Khudozhestvennaia literatura, 1983 [1957].

Tsunts, Mikhail. *Rasskaz o Bol'shoi Volge*. Moscow: Sovetskaia Rossiia, 1964.

Turpin, Trevor. *Dam*. London: Reaktion Books, 2008.

Tvardovskii, Aleksandr. *Za dal'iu dal'*. Moscow: Sovetskii pisatel', 1961.

Ulrich, Roberta. *Empty Dams: Indians, Dams, and the Columbia River*. Corvallis: Oregon State University Press, 1999.

Van Slyke, Lyman P. *Yangtze: Nature, History, and the River*. Reading, MA: Addison-Wesley, 1988.

Verhoeven, Harry. *Water, Civilization, and Power in Sudan: The Political Economy of Military-Islamist State Building*. Cambridge: Cambridge University Press, 2015.

Waterbury, John. *Hydropolitics of the Nile Valley*. Syracuse, NY: Syracuse University Press, 1979.

Waters, Frank. *The Colorado*. Athens, OH: Swallow, 1984 [1946].

Weiss, Thomas G., Tatiana Carayannis, Louis Emmerij, and Richard Jolly, eds. *UN Voices: The Struggle for Development and Social Justice*. Bloomington: Indiana University Press, 2005.

White, Richard. *The Organic Machine*. New York: Hill and Wang, 1995.

Wilson, Edmund. "Hoover Dam." In *The American Earthquake: A Documentary of the Twenties and Thirties*, by Wilson, 368–78. Garden City, NY: Doubleday, 1958.

Winchester, Simon. *The River at the Center of the World: A Journey up the Yangtze and Back in Chinese Time*. New York: Picador, 1996.

Xiaolong, Qiu. *Don't Cry, Tai Lake*. New York: Minotaur Books, 2012.

Ying, Hong. *Peacock Cries at the Three Gorges*. Translated by Mark Smith and Henry Zhao. London: Marion Boyars, 2004.

Zeisler-Vralsted, Dorothy. *Rivers, Memory, and Nation-Building: A History of the Volga and Mississippi Rivers*. New York: Berghahn, 2015.

Ziolkowski, Margaret. *Rivers in Russian Literature*. Newark: University of Delaware Press, 2020.

Index

Abbey, Edward, 167, 171; on Glen Canyon Dam, 160–61; *The Monkey Wrench Gang*, 162–64

Abu Simbel, 19, 93, 96, 100

Acts of War (Rovin), 156

Adivasi, 18, 101–2

Africa, 25

agriculture, 10, 93, 128

AIM. *See* American Indian Movement

Alabama, 15

Ali, Idris, 71, 94, 112, 175; *Dongola: A Novel of Nubia*, 95–97, 113, 127; on environmental damage, 114, 126–28

Amazon River, Amazonia, 9, 17

American Indian Movement (AIM), 89, 91

Amte, Baba, 102–3, 173

Angara (Taurin), 42; themes in, 43–45, 46

Angara-reka (The Angara River; Pristavkin), 46

Angara River, 25, 71, 172–73, 175; environmental damage, 114, 117–18; hydroelectric dams, 9, 41–42, 45; novels on, 42–43

anti-communism: and Boulder Dam Project, 54, 56–57; as novel theme, 54, 56–57

Appalachians, 17

archaeological sites, 19, 109; Aswan High Dam, 92–93

Arikara, 17

Armstrong, Dan, *Taming the Dragon*, 133, 135–37, 138–39

Around the Union of Soviets (Gorky), 31; on Volga, 38–39

Aswan! (Heim), 153–54

Aswan Assignment, The (Sugar), 153

Aswan Hellbox (Wilson), 153

Aswan High (Harrington and Young), 154, 155–56

Aswan High Dam, 5, 13, 19, 71, 91; environmental damage, 126–28; fictional attacks on, 152–56; Nubian relocation and, 92–93, 95–100

Aswan Low Dam, 13, 25, 71, 91, 98

Aswan Solution, The (Rowe), 154–55

Atatürk, Mustafa Kemal, 156

202 INDEX

Ataturk Dam, 156–57
Atbara River, 92
atom bomb, and Hanford plant, 62
Aurelia: A Crow Creek Trilogy (Cook-Lynn): environmental damage, 124–26; themes in, 87–91, 113

Baikal, Lake, 9, 41, 42, 45
Baikal Hydroelectric Station, 42
"Ballad of the Great Grand Coulee" (Guthrie), 61
ballads, Woody Guthrie's, 59, 60–61
Banqiao Dam, collapse of, 109, 146–47
Barber, Katrine, 19, 81
Barber, Phyllis, *And the Desert Shall Blossom*, 57
Bash, Iakov (Iakiv Bashmak), *Warm Feelings*, 35–36
Before the Deluge (Chetham), 137–38
Bezymenskii, Aleksandr, "A Tragic Night," 32–34
Big Bend Dam, 87
"Biggest Thing That Man Has Ever Done (the Great Historical Bum), The" (Guthrie), 61
Big Red (Haase), 51; themes in, 55–57
Big Volga Project, 39–40, 41
bilharzia, 114, 127
Black Canyon, 53; Hoover Dam, 48–49
Blow, The (Panferov), 40
Bolsheviks, 29
Bonneville Dam, 62, 81; construction of, 58–59
Bonneville Power Administration (BPA), Guthrie and, 59, 60–61
Boulder Canyon Project Act, 49, 50; novels on, 51–57
Boulder Dam, 3, 48. *See also* Hoover Dam
Boulder Dam (Grey), 28; themes in, 51–55
Bourassa, Robert, 7
BPA. *See* Bonneville Power Administration
Bratskaia GES (Evtushenko), 28, 46–47
Bratsk Dam, 5, 27, 172; in poetry and prose, 45–47; relocation in, 75–76
Brazil, 9, 17, 146
Brezhnev, Leonid, 172

Brower, David Ross, 23; on Glen Canyon Dam, 159–60
Brumadinho Dam, 146
Bruski (Panferov), 40
Buckles, Eleanor, *Valley of Power*, 64–66
Bureau of Indian Affairs, agency towns, 90–91

Cairo, Nubians in, 96, 97–98, 128, 175
Calahan, James M., 162
Cambor, Kathleen, *In Sunlight, in a Beautiful Garden*, 147, 148
Canada, First Nations relocation, 17
canals, Volga River, 39
Can You Survive the Johnstown Flood: An Interactive History Adventure (Otfinoski) 147
capitalism, 28; as victor, 54–55
Carnegie, Andrew, 7
caste system, Hindu Indian, 101
Celilo Falls, 81; in Craig Lesley's novels, 82–84
Celilo Village, 82
Cement (Gladkov), 34
cemeteries, 18, 65
Chelovek i oruzhie (Honchar), 36
Cheops' Pyramid, comparisons to, 13, 46–47
Chetham, Deirdre, *Before the Deluge*, 137–38
Cheyenne River Reservation, 87
China, 9, 12, 20, 25, 26, 175; Banqiao Dam failure in, 146–47; Three Gorges Dam, 5, 24, 108–13
Chongqing, and Three Gorges Dam, 137
Chudakova, Marietta, *"Plyvushchii korabl',"* 120
Churchill, Winston, 7
Circle of Dancers (Cook-Lynn), 87, 89–90
Clancy, Tom, 156
Cold War, 8–9, 10, 15, 26, 27, 48, 108, 153, 157; US-Soviet competition, 39–40, 68
Colorado, The (Waters), 57
Colorado River, 25, 48, 52, 159, 167; conquering, 53, 56
Columbia, The (film), 59
Columbia River Basin, 19, 25, 62; development of, 57–59; fisheries, 81–86, 114, 121–24, 168–71; Native Americans and, 21–22, 80–81

Columbia River Basin Project, and Woody Guthrie, 59–62
Colville Indians, 81
Concrete Revolution (Sneddon), 117
conservation, conservationists, 6, 23, 158–59
construction defects, Chinese dams, 109
"Conversation with Padun, A" (Tvardovskii), 45
Cook, Ian G., *Green China*, 109
Cook-Lynn, Elizabeth, 114, 175; *Aurelia: A Crow Creek Trilogy*, 87–91, 124–26
Coopey, Judith Redline, *Waterproof: A Novel of the Johnstown Flood*, 147–48, 151
corruption, 5, 17, 72, 96, 109
cost-benefit calculations, 72–73
Cracked: The Future of Dams in a Hot, Chaotic World (Hawley), 174
Cree, 17
Crow Creek Reservation, novels set on, 87–91
Crowe, Frank, 49; as novel character, 55–56
Cullen, Allen H., 8; *Rivers in Harness*, 9
cultural sites, 18, 19, 109; Aswan High Dam, 92–93. *See also* displacement
cyanobacteria, 157

Dai Qing, 133; novels by, 109
Dalits, 18, 101, 104
Dalles Dam, 81, 82
Dam Builders, The (Gulick), salmon fisheries in, 122–24
Dam Busters, The (film), 147
dam industry, 70–71
Daninos, Adrian, 92
Deal, Borden, *Dunbar's Cove*, 64, 66–68
"Decay of Lying, The" (Wilde), 6
decommissioning, dam, 26, 174
deforestation, 21, 131
deltas, 23, 167
derevenskaia proza, 76
Desert Shall Blossom, And the (Barber), 57
destruction, of dams, 147
development, 3, 5, 10, 11, 15, 69, 173; in India, 100–101
Dinosaur National Monument, 23, 159
Dirge for the Damned, A (Patil), themes in, 103–7, 112, 113, 129–31

disadvantaged populations, 5, 17–18; in India, 18, 101, 104. *See also* Native Americans; Nubians; relocation, resettlement
discrimination, in Boulder Canyon Project, 50
diseases, 22, 127, 175
displacement: impacts of, 70–71, 75, 112–13; of Indian populations, 101–2, 103–7; Three Gorges Dam, 111–12
Dneprostroi Dam (DneproGES), 5, 12, 15, 27, 29, 62; destruction and reconstruction of, 36–38; Gorky and, 30–31; novels on, 34–36; paintings on, 31–32; poetry on, 32–34
Dnieper River, 25, 28–29, 36
"Dolg Ukraine" (Maiakovskii), 32
Dongola: A Novel of Nubia (Ali), themes in, 95–97, 113, 127
Don't Cry, Tai Lake (Xiaolong), 143
Donzère-Mondragon Dam, 13
"Downstream and Upstream" (Rasputin), 76, 80, 144; environmental themes in, 118–20
"Down the River" (Abbey), 161
Drobotov, Viktor, 120
D'Souza, Dilip, 102
Dujiangyang irrigation system, 140–41
Dunbar's Cove (Deal), 64; themes in, 66–68
"Duty to Ukraine" (Maiakovskii), 32

Earth First!, 163
earthquakes, 22, 131, 132, 158
Echo Park Dam, 23, 159
ecological impacts, reservoirs, 21–23
economic planning, development, 12, 74
ecosystems, impacts on, 21–23
ecoterrorism, ecotage, 26, 162; as novel theme, 163–71
Eco-Warriors, Nihilistic Terrorists, and the Environment (Likar), 161–62
Eco-warriors: Understanding the Radical Environmental Movement, 162
Edersee Dam, destruction of, 147, 153
Egypt, 13, 18, 26; Aswan High Dam, 5, 91–92, 93–95
Ekbladh, David, 10
Ekologicheskaia al'ternativa (The Ecological Alternative), 121

204 INDEX

electrification, 10, 15
embankment dams, 14
employment, dams as symbols of, 72
Energiia (Energy) (Gladkov), 28, 34–35
engineering, engineers, 14, 30, 33
Enlightenment, 7, 10
environmental damage, 5, 26, 72, 114–15,
  173, 174; Angara River reservoirs, 117–20;
  Aswan dams, 93–94, 126–28; Columbia
  River dams, 121–24; India, 129–31; Missouri River dams, 124–26; Three Gorges
  Dam, 109; Volga River reservoirs, 120–21
environmental movement, 161–62, 163
erosion, 21, 131
essays, on Russian rivers, 120–21
Euphrates River, Ataturk Dam and, 156
evaporation, 22
Everard, Mark, 75, 115
Evtushenko, Evgenii, *Bratskaia GES*, 28, 46–47

failures, dam, 109, 145–49, 157–58, 173
Farewell to Matyora (Rasputin), 76, 80, 175;
  relocation themes in, 77–79, 112, 113
female abandonment, Nubian society, 94,
  97, 99
*Fifth Horseman: A Novel of Biological Disaster,
  The* (Sherbaniuk), 156, 157
films, 16, 59, 64
"Fire, The" (Rasputin), 76, 79, 80
Fires of Communism, The (newspaper), 43
First Nations, relocation of, 17
fisheries, Columbia River, 62, 80–82, 83–86,
  121–23, 169–71
Five-Year Plan, Dneprostroi, 31
flood control, 13, 14, 48
floods, flooding, 23, 36, 49, 98, 108, 122, 126,
  132, 140; from dam failures, 145–46
Fort Randall Dam, 87
France, 9, 13
Fulton, John, 148

Georgia, 15
Germany, 6, 147
Gezhouba dam, 109
gigantism, 173
Gladkov, Fedor, 28; novels by, 28, 34–35

Glen Canyon, 23
Glen Canyon Dam, 5, 165–67; Edward Abbey
  on, 160–61, 162–64; David Brower on,
  159–60
Global South, 12
global warming, 21
GOELRO (State Electrification of Russia), 29
Goldsmith, Edward, 72, 74
Gorky, Maksim, 30, 42, 61; and Dneprostroi,
  31–32; on Volga, 38–39
Grace, Stephen, 13
Grand Coulee Dam, 5, 13, 14, 27, *58f*, 168;
  comparisons to, 141–42; salmon fisheries,
  62–63, 80–81; and Woody Guthrie, 59–60,
  61, 172
graves, 18, 65
gravity dams, masonry, 14
Great Britain, and Aswan High Dam, 92
Great Depression, 50, 58–59
"Greater Common Good, The" (Roy), 102
Great Flood (1993), 126
Great Marib Dam, 145
Great Pyramids, 13
Green China: Seeking Ecological Alternatives
  (Murray and Cook), 109
Greener, Leslie, 19–20
greenhouse gases, reservoirs and, 21
Green River, 23
Grey, Zane, 27–28; *Boulder Dam*, 38, 51–55
Gujarat, 102, 146
Gulick, Bill, 114; *Roll on, Columbia* trilogy,
  122–24
Gunther, John, *Inside U.S.A.*, 51, 62–63, 68
Guthrie, Woody, 28, *59f*; and Columbia River
  Basin Project, 28, 59–62, 172
"Guys on the Grand Coulee Dam, The"
  (Guthrie), 61
Gwalior, 146

Haase, John, *Big Red*, 51, 55–57
Hamdi, Abdelrahim, 64
Hanford site, 62
Hansen, Gary, *Wet Desert: Tracking down a
  Terrorist on the Colorado River*, 164–67
Hariachi pochuttia (Bashmak), 35
Harrington, R. E., *Aswan High*, 154, 155–56

Index   205

Harris, Thomas V., *Three Gorges Dam*, 158
Hawley, Steven, 11; *Cracked*, 174
Heim, Michael, *Aswan!*, 153–54
Hersey, John, *A Single Pebble*, 133–35
Hetch Hetchy dam, 159
Hidatsa, 17
Hilyard, Nicholas, 72, 74
Hindu Indian culture, caste system, 101–2
Hittmaier, Otto, 8, 10
Hockenberry, John, *A River Out of Eden*,
    169–71
Honchar, Oles', *Liudyna i zbroia*, 36–38
Hong River, 109
Hoover, Herbert, 48
Hoover Dam, 3, 4f, 14, 24, 27, 48–49f, 50, 172,
    173; novels on, 51–57
hydroelectric projects, 5, 6, 48
hydropower, 15–16, 62, 117

ICOLD. *See* International Commission on
    Large Dams
Idaho, 146
illness, from relocation, 75
India, 5, 13, 16, 24, 25, 26, 146, 173; cultural/
    social structure, 101–2; environmental
    damage, 129–31; Five Year Plan, 100–101;
    relocations in, 17–18, 20, 21, 103–7;
    remediation of nature, 6–7; resistance/
    opposition in, 102–3
Indigenous peoples: treatment of, 17–20. *See
    also* Native Americans; Nubians
industrialization, 5, 10
Industrial Workers of the World (IWW),
    50, 56
Inside U.S.A. (Gunther), 51, 68; on Columbia
    River salmon industry, 62–63
interactive games, on Johnstown Flood, 147
International Commission on Large Dams
    (ICOLD), 8, 16
Into the Desert (Gulick), 122
Iraq, and Ataturk Dam, 156
Irkutsk Dam, 42–43, 46
Irkutsk Hydroelectric Station, 42
irrigation, 13, 48, 93; Dujiangyan project,
    140–41
Israelis, in Aswan Dam novels, 151–56

Italy, Vajont Dam, 146
Ivankovo, 39
IWW. *See* Industrial Workers of the World

Jackson, Donald C., *Pastoral and Monumen-
    tal*, 5
Jambhli Dam Project, 104–5
James Bay, 17
Jhadajhadati (Patil), 103. *See also Dirge for the
    Damned, A*
Jog Falls, 7
Johnstown, and South Fork Dam, 148, 149
Johnstown Flood, 145–46; events of, 148–49;
    novels about, 147–48, 149–51
*Johnstown Flood: An UP2U Historical Fiction
    Adventure, The* (Mullarkey and Mullar-
    key), 147
Josephson, Paul R., 9, 12, 16

Kahn, Stephen, 59, 60
Karababa Dam. *See* Ataturk Dam
Kazan, Elia, 64
Kentucky, 15
Kettle Falls, 168
K Falls (Moody), 167; ecoterrorism theme
    in, 168–69
Khagram, Sanjeev, 11
Khashm al-Girba, 92
Khortytsia Island, 30
Khrushchev, Nikita, 8, 93f
Klingensmith, Daniel, 30
Kom Ombo Valley, 92
Kotov, Pyotr Ivanovich, *Maxim Gorky at the
    Construction of the Dnieper Hydroelectric
    Station*, 31–32, 33f
Koyna Hydroelectric Project, 104
Kuo, Alex, 114; *The Man Who Dammed the
    Yangtze*, 133, 139–43

labor migration, Nubian, 94, 97, 99, 100
Lap of the Gods, In the (Lovett), themes in,
    110–13
Lawson, Michael L., 86
Lenin, Vladimir, 10, 29
Lesley, Craig, 82, 124, 175; *River Song*, 84–86,
    121–22; *Winterkill*, 83–84

206 INDEX

Libya, dam failure in, 146
Lieberman, Avigdor, 152
Likar, Lawrence E., *Eco-Warriors, Nihilistic Terrorists, and the Environment*, 161–62
Lilienthal, David E., 12; *TVA: Democracy on the March*, 7–8, 10, 14–15, 63
Little Mother Volga (Panferov), 40–41
Liudyna i zbroia (Honchar), 36–38
Lomax, Alan, 59, 60–61
Los Angeles, St. Francis Dam collapse, 146
Lovett, Li Miao, 71, 175; *In the Lap of the Gods*, 110–13
Lower Brule Reservation, 87

Machchu-2 Dam, 146
Madhya Pradesh, 102, 146
Mahar, 104
Maharashtra, 102
Maiakovskii, Vladimir, "Duty to Ukraine," 32
malaria, 22
Mandan, 17
Manhattan Project, 62
Man Who Dammed the Yangtze, The (Kuo), 133; themes in, 139–43
Mao Zedong, 12; and Three Gorges Dam, 108–9
Matyora, 77, 175
Mauvoisin Dam, 14
Maxim Gorky at the Construction of the Dnieper Hydroelectric Station (Kotov), 31–32, 33f
McCullough, David, 149
McCully, Patrick, 20, 23, 131; on resettlement, 74, 75; *Silenced Rivers*, 16–17
McNeill, J. R., 19
Mead, Lake, seismicity, 132
Meditation (Panferov), 40
Mediterranean, 23
memoir, *Angara-reka*, 46
Minas Gerais, Brumadinho Dam in, 146
minority populations, relocation of, 17–18
Min River, 140
Mississippi, 15
Mississippi River, 6, 10, 49
Missouri River, 25; environmental damage, 114, 124–26; Native Americans on, 86–91

Moans of the Volga, 120–21
modernism, modernity, 3, 9–10, 72; high, 38, 52–53, 63–65, 67, 69
Möhne Dam, 147
Monkey Wrench Gang, The (Abbey), 162–64
Moody, Skye Kathleen, *K Falls*, 167–69
Morbi, Machchu-2 Dam, 146
Morrell, Daniel J., 148
mortality, relocation, 75
Mother, Volga as, 39, 120–21
Mother India (film), 16
Muir, John, 159
Mulholland, William, 146
Mullarkey, John and Lisa, *The Johnstown Flood*, 147
multi-use dams, 14
Mumford, Lewis, 7
Murray, Geoffrey, *Green China*, 109

Nagasaki, atom bomb, 62
Name of the Young, In the (Panferov), 40
Narmada Bachao, 24
Narmada River Project, 24; environmental damage, 129–31; opposition, 102–3
Nasser, Gamal Abdel, 152; and Aswan High Dam project, 12–13, 92, 93f
Nasser, Lake, 93
nationalism, 4, 10
Native Americans, 17, 175; and Columbia River Basin, 21–22, 80–86; and Missouri River dams, 86–91, 124–26; and salmon river fisheries, 121–24
nature, 6, 35, 52, 159; human domination of, 7–9, 24–25; victory over, 34, 53–54, 64–65
navigation, 6, 13, 48
Nehru, Jawaharlal, 13, 20, 24, 172, 173
New Celilo, 82
New Deal, 12, 15, 59, 62, 63
New Halfa, 92
New Nubia, 92, 94; as novel theme, 95–96
Nez Perce, 80, 81; in Craig Lesley's novels, 82–86
Nights of Musk: Stories from Old Nubia (Oddoul), themes in, 98–100, 128
Nile River, 23, 175; environmental damage, 114, 126–28

Index 207

North Carolina, 15
North Dakota, 17, 87
Northern Sudan, 64
"Notes of My Contemporary" (Pristavkin), 46
novels: on Angara River projects, 42–43; on Ataturk Dam, 156–57; of attack on Aswan Dam, 152–56; on Boulder Dam project, 51–56; on Celilo Falls, 82–86; on Columbia River, 122–24; on Dneprostroi, 34–36; ecoterrorism themes in, 162–71; on Indian projects, 103–7; on Johnstown Flood, 147–52; on Missouri River dams, 87–91; on Nubian relocation, 95–100; relocation themes in, 112–13; on Tennessee Valley Authority, 64–65; on Three Gorges Dam, 109–12, 133–43, 157–58
Nubia, 19, 71, 92–93
Nubians, 125; and Nile River damage, 126–28; relocation of, 18, 92–93, 94–100

Oahe Dam, 87, 88
Oddoul, Haggag Hassan, 94, 126, 175; Nights of Musk, 98–100, 128
Operation Chastise, 147
opposition: in India, 102–3; to Three Gorges dam, 109
Oregon, 58
Osborn, Paul, Wild River, 64
Otfinoski, Stephen, Can You Survive the Johnstown Flood, 147
Owens Falls, 7

Padun Rapids, 45
paintings, on Dneprostroi, 31–32f
Pal'kin, Nikolai, 121
Palmer, Tim, 13
Panferov, Fedor, Volga matushka reka, 40–41
Panshet Dam, collapse of, 146
Parman, Donald L., 87
Pastoral and Monumental: Dams, Postcards, and the American Landscape (Jackson), 5
Patil, Vishwas, A Dirge for the Damned, 103–7, 112, 129–31
Patkar, Medha, 24, 102
Pearce, Fred, 19
Pennsylvania, Johnstown Flood, 145–46

People and Arms (Honchar), 36–38
Peshkov, Aleksei. See Gorky, Maksim
Pick, Lewis A., 86
Pick-Sloan Missouri Basin Program, 86–87; farm and ranch program, 89–90
Pieczenik, Steve, 156
Pittsburgh, and South Fork Hunting and Fishing Club, 149
planning process, rural populations and, 71–72
"Plyvushchii korabl'" (Chudakova), 120
poetry, 45; on Bratsk project, 46–47; on Dneprostroi, 32–34
politics, 3–4, 17; Columbia River Basin, 57–58
pollution, 127, 173, 175; Angara River reservoirs, 118, 119; Yangtze River, 137–39, 143
poor, displacement of, 19, 21
Poor (Ali), 95, 97–98, 113, 127–28
Po Soiuzu Sovetov (Gorky), 31; on Volga, 38–39
Powell, Lake, 160
"Pozhar" (Rasputin), 76, 79, 80
Presence of River Gods, In the (Cook-Lynn), 87, 90–91, 125–26
Pristavkin, Anatolii, 172–73; Angara-reka, 46
propaganda: Aswan High Dam as, 12–13; Three Gorges Dam as, 108–9; TVA, 63–64
"Proshchanie s Materoi" (Rasputin), 76, 80, 175; relocation themes in, 77–79, 112, 113
protests, 24; against Three Gorges dam, 109, 112
public consciousness, 5
Public Works Administration, 58–59
Pune, 146
pyramids, dam comparisons to, 13, 46–47, 52, 53

Qingcheng Shan, 140
Quebec, 7
Qutang Gorge, 108

racism, 90–91, 94
RAF Bomber Command, dam destruction, 147
Ragtown, 50
Rasputin, Valentin, 71, 75, 112; "Downstream and Upstream," 76, 80, 118–20, 144; envi-

208 INDEX

ronmental themes, 114, 117–18; "Farewell to Matyora," 77–79, 175; "The Fire," 79
Razdum'e (Panferov), 40
"*Razgovor s Padunom*" (Tvardovskii), 45
reclamation, rectification, 6
Red River, floods, 126
Red River (Tonkin), 157–58
regional planning agencies, 15
Reisner, Marc, 13
relocation, resettlement, 2; Bratsk Dam, 75–76; impacts of, 17–20, 73–74, 76–80, 87, 173–74; in India, 101–2, 103–6; Nubians, 92–93, 94–100; themes of, 112–13; Three Gorges dam, 109–12
reservoirs, 13; environmental damage from, 21–22, 118–20, 125–26, 129–32, 173
resistance, 24, 88; in India, 102–3, 106–7
Rhine River, rectification of, 6, 9
Rhône River, 9
Rich, Bruce, 20–21, 74
Rist, Gilbert, 11
River, The (film), 59
river basin planning, 14
River Dragon Has Come!, The (Dai Qing), 109
River out of Eden, A (Hockenberry), 169–71
River's Edge, From the (Cook-Lynn), 87–88
Rivers in Harness (Cullen), 9
River Song (Lesley), 82, 84–86; on salmon fisheries, 121–22, 124
"Roll, Columbia, Roll" (Guthrie), 61
"Roll On, Columbia, Roll On" (Guthrie), 28, 61
Roosevelt, Franklin Delano, 3, 4f, 7, 63, 172
Roosevelt, Lake, 81, 168
Rovin, Jeff, *Acts of War*, 156
Rowe, John, *The Aswan Solution*, 154–55
Roy, Arundhati, 173; "The Greater Common Good," 102
Rubin, Neville, 73
Ruff, Benjamin, 148
Ruhr River, 147
rural populations: in India, 101–2; and planning process, 71–72; relocation, 17, 73–74
Ru River, 109; Banqiao Dam, 146–47
Russia, 5, 16, 76, 144
Rybinsk, 39

"Sailing Ship" (Chudakova), 120
St. Francis dam, collapse of, 146
salmon runs, fisheries, 19; Columbia River Basin, 21, 62–63, 80–82, 114, 121–24, 168–71
Sandouping, 108
San Francisquito Canyon, 146
sardine catch, Mediterranean, 23
Savage, John L., 108
Scarce, Rik, *Eco-warriors*, 162
Scheduled Castes, 101, 104
Scheduled Tribes, 101–2
schistosomiasis, 22, 93
Scott, James C., on high modernism, 9–10
scouring, 23
Scudder, Thayer, 20, 74
Second Arab Israeli War, 152
sedimentation: in reservoirs, 131–32; Three Gorges Dam, 132–33, 136–37, 141
seismicity, 132. *See also* earthquakes
settler colonialism, 7, 12, 69, 88, 90, 92
Shapiro, Judith, 9
Sherbaniuk, Richard, *The Fifth Horseman*, 156, 157
Shimantan Dam, 109
Siberia, 9, 16, 17, 25, 26, 76; Angara River projects, 41–42
Sierra Club, 23
Silenced Rivers: The Ecology and Politics of Large Dams (McCully), 16–17
siltation, 22
Sinedubsky, Vladimir, 9
Single Pebble, A (Hersey), on Three Gorges Dam, 133–35
Sioux, and Missouri River dams, 87–91
Six-Day War, 152
Slavutich, 36; Volga as, 38–39
Sloan, William Glenn, 86
Sneddon, Christopher, *Concrete Revolution*, 117
social impacts, 23–24; of relocation, 73–74, 75, 76–80, 88–91, 95–100
socialism, 15, 28, 48
social transformation, dams as, 60
"Song of the Grand Coulee Dam, The" (Guthrie), 61
songs, Woody Guthrie's, 59–62

South Dakota, 87

South Fork Dam, Johnstown Flood and, 145–46, 148–49

South Fork Hunting and Fishing Club; and Johnstown Flood, 148, 149, 150–51

Soviet Union, 5, 11, 12, 17, 25, 92, 108; Cold War, 8, 9, 26, 27, 39–40, 68; Dneprostroi, 29–38; hydropower competition, 15–16; and Volga, 38–39

species diversity, 21

Spengler, Oswald, 7

Spokane Indians, 81

Standing Rock Reservation, 87

State Electrification of Russia (GOELRO), 29

status symbols, dams as, 4

Stony Volgi, 120–21

stupas, comparisons to, 13

subsistence economy, 82; India, 101; Sioux, 87, 89–90

Sudan, 64, 92

Suez Canal, 152

Sugar, Andrew, The Aswan Assignment, 153

Sunlight, in a Beautiful Garden, In (Cambor), 147, 148; plot of, 149–51

Sun Yat Sen, 108

symbolism, of dams, 13

Syria, and Ataturk Dam, 156

Taming the Dragon (Armstrong), 133; themes in, 135–37, 138–39

Taurin, Frants, Angara, 42–45, 46

technology, 12, 16; and Three Gorges Dam, 134–35

temples, comparisons to, 13

Tennessee, 15

Tennessee River, controlling of, 67, 68

Tennessee Valley, 15, 25

Tennessee Valley Authority (TVA), 17, 27; novels on, 64–68; river basin planning, 14–15, 63–64; symbolism of, 39–40

Terkel, Studs, 60

terrorism, 5, 26, 162. See also ecoterrorism, ecotage

Teton Dam, collapse of, 146

Three Gorges Dam, 5, 24, 25, 71, 140–44; ambivalence about, 134–36; catastrophic

topics, 157–58; criticism of, 109–13; Mao Zedong and, 108–9; pollution, 137–39; sedimentation, 132–33, 136–37

Three Gorges Dam (Harris), 158

Tigra Dam, collapse of, 146

Toc, Mount, 146

Tonkin, Peter, Red River, 157–58

Tragediinaia noch'" (Bezymenskii), 32–34

treaty rights, Columbia River Basin, 80, 81

Trotsky, Leon, 29

Truman, Harry, on Columbia River Dam, 62

Tsement (Gladkov), 34

Tsukanov, Aleksandr, 120

Turkey, Ataturk Dam, 156–57

TVA. See Tennessee Valley Authority

TVA: Democracy on the March (Lilienthal), 7–8, 14–15, 63

Tvardovskii, Aleksandr, "Razgovor s Padunom," 45

Typhoon Nina, and Banqiao Dam, 146–47

Udar (Panferov), 40

Uglich, 39

Ukraine, 5; Dnieper River, 25, 28–29, 36–38

Ulrich, Roberta, on Celilo Falls, 81

Umatilla, 81, 83

United Nations, 11–12

United States, 5, 16, 24, 25, 30, 48, 92; Cold War, 8, 26, 27, 39–40, 68; dam failures, 145–46

US Army Corps of Engineers, 48, 58, 86

US Bureau of Reclamation, 6, 48, 58, 86, 108

US Congress, 19, 63, 87

US Department of the Interior, 59, 60

US Export-Import Bank, 133

US Supreme Court, 80

utility companies, 48

Vajont Dam, collapse of, 146

Valley of Power (Buckles), themes in, 64–66

Vanport Flood, 122

Van Slyke, Lyman P., 134–35

victimization, argument against, 103–4

Victoria, Lake, 7

village/country prose, 76

Vinter, Aleksandr, 33

210 INDEX

Virginia, 15
Visvesvaraya, Mokshagundam, 6–7
"Vniz I verkh po techeniiu" (Rasputin), 76, 80, 144; environmental themes in, 118–20
Vo imia molodogo (Panferov), 40
Volga-Don Canal, 39
Volga matushka reka (Panferov), 40–41
Volga-Moscow Canal, 39
Volga River, 10, 16, 25, 175; environmental damage, 120–21; Panferov trilogy on, 40–41; Soviet projects on, 38–39

Wadi Halfa District, 92
Warm Feelings (Bash), 35–36
Warm Springs tribes, 81
Warren, William M., 73
Washington (state), 58
Waterbury, John, 12
Waterproof: A Novel of the Johnstown Flood (Coopey), 148, 151
water quality, 21, 22, 173
Waters, Frank, 24; The Colorado, 57; on Hoover Dam, 50–51, 172, 173
Wet Desert: Tracking down a Terrorist on the Colorado River (Hansen), ecoterrorism theme in, 164–67
Wilde, Oscar, "The Decay of Lying," 6
wilderness areas, 23

wildlife, reservoir impacts on, 130–31
Wild River (film), 64
Wilson, Edmund, 51
Wilson, Gar, Aswan Hellbox, 153
Winchester, Simon, 12
Winterkill (Lesley), themes in, 82, 83–84, 124
World Bank, 11–12, 19, 20–21, 24
World War II, 42, 62, 147; Dneprostroi destruction, 36–38
Wu Gorge, 108, 134

Xiaolong, Qiu, Don't Cry, Tai Lake, 143
Xiling Gorge, 108, 134

Yakama, 17, 81
Yangtze River, 25, 158; pollution, 137–39, 143; sedimentation on, 114, 141; Three Gorges Dam, 108–13
Yangtze, Yangtze (Dai Qing), 109
Yankton Reservation, 87
Yemen, Great Marib Dam in, 145
Yom Kippur War, 152
Yosemite National Park, 159
Young, James A., Aswan High, 154, 155–56

"Zapiski moego sovremennika" (Pristavkin), 46
Zaporizhzhia, 30
Zeisler-Vralsted, Dorothy, 10